Studying Mathem
Applicati

Palgrave Study Guides

A Handbook of Writing for Engineers *Joan van Emden*
Effective Communication for Science and Technology *Joan van Emden*
How to Write Better Essays *Bryan Greetham*
Key Concepts in Politics *Andrew Heywood*
Linguistic Terms and Concepts *Geoffrey Finch*
Literary Terms and Criticism (second edition) *John Peck and Martin Coyle*
The Mature Student's Guide to Writing *Jean Rose*
The Postgraduate Research Handbook *Gina Wisker*
Practical Criticism *John Peck and Martin Coyle*
Research Using IT *Hilary Coombes*
The Student's Guide to Writing *John Peck and Martin Coyle*
The Study Skills Handbook *Stella Cottrell*
Studying Economics *Brian Atkinson and Susan Johns*
Studying History (second edition) *Jeremy Black and Donald M. MacRaild*
Studying Mathematics and its Applications *Peter Kahn*
Studying Psychology *Andrew Stevenson*
Teaching Study Skills and Supporting Learning *Stella Cottrell*

How to Begin Studying English Literature (second edition) *Nicholas Marsh*
How to Study a Jane Austen Novel (second edition) *Vivien Jones*
How to Study Chaucer (second edition) *Rob Pope*
How to Study a Charles Dickens Novel *Keith Selby*
How to Study Foreign Languages *Marilyn Lewis*
How to Study an E. M. Forster Novel *Nigel Messenger*
How to Study a Thomas Hardy Novel *John Peck*
How to Study James Joyce *John Blades*
How to Study Linguistics *Geoffrey Finch*
How to Study Modern Drama *Tony Curtis*
How to Study Modern Poetry *Tony Curtis*
How to Study a Novel (second edition) *John Peck*
How to Study a Poet (second edition) *John Peck*
How to Study a Renaissance Play *Chris Coles*
How to Study Romantic Poetry (second edition) *Paul O'Flinn*
How to Study a Shakespeare Play (second edition) *John Peck and Martin Coyle*
How to Study Television *Keith Selby and Ron Cowdery*

www.palgravestudyguides.com

Studying Mathematics and its Applications

Peter Kahn

palgrave

© Peter Kahn 2001

All rights reserved. No reproduction, copy or transmission of this publication may be made without written permission.

No paragraph of this publication may be reproduced, copied or transmitted save with written permission or in accordance with the provisions of the Copyright, Designs and Patents Act 1988, or under the terms of any licence permitting limited copying issued by the Copyright Licensing Agency, 90 Tottenham Court Road, London W1T 4LP.

Any person who does any unauthorised act in relation to this publication may be liable to criminal prosecution and civil claims for damages.

The author has asserted his right to be identified as the author of this work in accordance with the Copyright, Designs and Patents Act 1988.

First published 2001 by
PALGRAVE
Houndmills, Basingstoke, Hampshire RG21 6XS and
175 Fifth Avenue, New York, N.Y. 10010
Companies and representatives throughout the world

PALGRAVE is the new global academic imprint of
St. Martin's Press LLC Scholarly and Reference Division and
Palgrave Publishers Ltd (formerly Macmillan Press Ltd).

ISBN 0–333–92279–4

This book is printed on paper suitable for recycling and made from fully managed and sustained forest sources.

A catalogue record for this book is available from the British Library.

10 9 8 7 6 5 4 3 2 1
10 09 08 07 06 05 04 03 02 01

Printed in China

Ad majorem dei gloriam

Contents

List of Figures	x
List of Examples	xii
Preface: How to Use this Book	xiii
Acknowledgements	xvi

1 **Setting the Scene**	1
What is mathematics?	1
Facts or connections?	2

Part I Skills

2 **Using Examples**	9
Why use examples?	9
Creating examples	12
Particular instances of examples	17
3 **Thinking Visually**	22
Why use visual images?	22
Drawing visual images	24
How to use visual images	27
4 **Coping with Symbols**	32
The purpose of symbols	32
Finding meaning in symbols	34
Developing fluency with symbols	37
5 **Taking Ideas Apart**	41
Why take ideas apart?	41
How to take an idea apart	43
Include all of the basic ideas	47
Revising basic ideas	49

6 Thinking Logically — 52
- Why think logically? — 52
- An introduction to logic — 56
- Further logic — 59

7 Making Connections — 63
- Connections in mathematics — 63
- How to make connections — 66
- Connections in specific areas of mathematics — 71

Part II Tasks

8 Solving Problems — 79
- Will any strategy do? — 79
- How to solve it — 80
- Case-studies — 88

9 Applying Mathematics — 96
- Why apply mathematics? — 96
- How to apply mathematics — 97
- Case-study: a model of a lever — 106

10 Constructing Proofs — 112
- The need for understanding — 112
- Understand the theorem — 113
- See the overall structure — 118
- See the details — 121
- Extension material: $\sqrt{2}$ is not a rational number — 123

Part III Study

11 Studying Actively — 131
- Introduction — 131
- Study involving tutors — 131
- Studying with other students — 136
- Independent study — 137

12 Using Technology — 140
- Introduction — 140
- Use the technology — 141

Employ the skills 142
Carry out the tasks 146
Case-study: use of a computer algebra system
in modelling population growth 148

13 Succeeding in Assessment 154
Focus on communication 154
Coursework 156
Examinations 162
Final word 164

Appendix: Mastering Algebra 166
Answers and Comments for Selected Exercises 170
Bibliography 185
Index 187

List of Figures

1.1	Some attitudes to mathematics and its applications	2
1.2	Approaches to studying mathematics and its applications	3
1.3	Two approaches to studying mathematics and its applications	4
2.1	Mathematical ideas can represent concepts from the real world	11
2.2	A set of examples of the real-world concept of distance and a matching set of real numbers	11
2.3	Graphs for the rules f and g	14
2.4	Graph of speed against time	19
3.1	A double helix	22
3.2	A circle	23
3.3	A tangram	27
4.1	A road sign	33
5.1	Dust disk around a massive black hole	42
6.1	Calculating π	53
6.2	The angles of a triangle	54
8.1	A structured approach to problem-solving	87
9.1	A typical portion of a street map	97
9.2	Mathematical ideas represent real-world concepts	98
9.3	The modelling cycle	99
9.4	A graph of demand against price	102
9.5	A model of a lever	107
9.6	Models of a lever indicating two fundamental assumptions	108
9.7	Models of a lever in equilibrium	109
10.1	A collection of examples of the theorem	114
10.2	The graph of the real-valued function given by $f(x) = 1 - x^2$	115
10.3	$\sqrt{2}$ as the length of a hypotenuse	124

10.4	Proof by contradiction	126
12.1	The graph of $P(t) = 3e^{2\sin t}$	149
12.2	Vectors representing selected values of *dP/dt* at different points of the plane $(t, P(t))$, when $a = 2$	150
12.3	A graph of the function $P(t)$, when $a = 2$ and $c = 3$, superimposed upon vectors representing selected values of *dP/dt* at different points of the plane $(t, P(t))$, when $a = 2$	151
12.4	Graphs of the functions $P(t)$, when $a = 2$, $c = 5$ and when $a = 2$, $c = 2$, superimposed upon vectors representing selected values of *dP/dt* at different points of the plane $(t, P(t))$, when $a = 2$	151
13.1	Characteristics of effective communication	154
A.1	A strategy to develop your fluency at mental algebra	166

List of Examples

2.1	Modelling the concept of distance	10
2.2	The concept of the speed of a body	15
2.3	Modelling the speed of a car	19
3.1	A ladder leaning against a wall	24
3.2	Sketching the graph of a function	25
3.3	Viewing the whole of an image and viewing the details	28
3.4	Connecting an image with formal mathematical ideas	29
3.5	Connecting an image with the real world	29
4.1	The meaning of the symbol '='	35
4.2	Translation of a text	37
4.3	Translation of another text	37
4.4	Partial translation of a text	38
5.1	Take apart the idea of an algebraic equation	43
5.2	Take apart the ideas of a statement, the number two, an algebraic equation and equality	44
5.3	Analysis of Newton's Second Law of Motion	46
6.1	Inductive reasoning in mathematics	55
6.2	A deduction	58
6.3	Implication or equivalence?	61
7.1	Rule to add two fractions together	65
7.2	Generalisation of a pairing of numbers	68
A.1	Mental solution of a linear equation	167
A.2	Mental rearrangement of an equation	168

Preface: How to Use this Book

This book is designed to help you study mathematics and its applications as effectively as possible. If your study focuses largely, or even exclusively, on mathematics or if you are studying one of the sciences, engineering, business studies, economics, statistics, computer science or any other subject involving the application of mathematics, then this is the book for you. It introduces you to the skills and strategies that you need in order to succeed.

The primary audience for this book is undergraduate students. The large majority of the material in this book will be accessible to students studying subjects which involve the application of mathematics. But additional extension material, which is clearly indicated, has also been provided in order to challenge students whose study focuses on mathematics itself. The book, however, will also be useful to any student whose study of mathematics or one of its applications involves more sophisticated mathematical ideas than just those of elementary algebra.

We will focus here exclusively on the distinctive approaches needed to study mathematics and its applications. What you will not find here is more general advice on time management, positive thinking and so on. To study mathematics and its applications effectively you need specialist advice.

The introductory chapter describes the approach to study that the book recommends. It aims to convince you that there is a better way of studying mathematics and its applications than endless practice and memorising formulae. Then in Part I of the book we take a look at six skills that are aimed at helping you make sense of mathematics and its applications. These are the skills of using examples of ideas, thinking visually, coping with symbols, taking ideas apart, thinking logically and making connections between ideas. Each of these chapters makes extensive use of illustrations from applications of mathematics as well as from mathematics itself.

In Part II we employ each of these skills to tackle the main tasks that you are likely to face in your study: solving problems, applying mathematics and establishing the truth of mathematical statements.

Finally, in Part III we see how you can make the best use of the skills and strategies covered earlier in the book in the context of your own study and assessment, with particular attention to the use of technology. The appendix to the book then provides some advice on developing a fluent grasp of algebra. The diagram summarises the approach that the book takes:

Part I
Introducing the skills needed to make
sense of mathematics and its applications.

Part II
A focus on the tasks you will meet in your study.

Part III
Strategies to face the practical challenges
of your study and assessment.

The recommended way to approach the book is to work through from beginning to end, and then to return to specific chapters which interest you. If you want to, however, you can approach the book more flexibly. For instance, if you are particularly interested in applications of mathematics you could focus your attention on Chapter 9 which covers how to apply mathematics (although ideally only after working through Part I of the book).

It is also worth observing that while this book is focused on developing study skills, it does cover a number of mathematical ideas and applications. In particular, both the examples in the text and the exercises draw on mathematical ideas and applications with which you are already likely to be familiar. They have been set at this level so that you can concentrate on developing study skills rather than on mastering new mathematical ideas. But having said this, some of you may find it helpful to revise any ideas you are less familiar with.

Finally, it is worth pointing out a couple of features that are common to the chapters which follow. Each chapter contains several exercises for you to work through. They are there for a reason: unless you actively try out the skills for yourself you are unlikely to learn them. Answers and comments on the exercises are provided at the back to help guide your initial work on these skills, and each chapter ends with some prompts for you to think about how effectively you are learning the skills. They are designed to help you take charge of your own learning.

This is not a book that tells you how to fit in that crucial extra 11 minutes of study by surfing a mathematics web-site on your mobile phone as you journey home after your lectures. My hope is that this book will encourage and help you to invest yourself in your study of mathematics and its applications.

Acknowledgements

I am grateful to my former colleagues at Liverpool Hope University College – in particular to Alf Westwell, Tony Fleet, John Brinkman and Jill Armstrong – for their support and to Liverpool Hope itself for the opportunity to engage in teaching experiences and research which helped lead to this book. I am further grateful to Celia Hoyles and David Tall for their encouragement in various ways. Comments were received from several individuals, including Keith Austin, Tony Barnard, Maria Charalambides, Robert Smedley and Mary Stevenson, and thanks are due to them.

Thanks are due to several students who commented on drafts of the book, including Peter Hartley, Anne Morris, Joseph Huang and Sarah Thomson. Numerous students at Liverpool Hope also contributed to research which helped to underpin this text and again I am grateful to these students.

I would like to acknowledge comments made by several reviewers. These comments have helped to shape the final text of the book. I have worked with various publishers at Palgrave and I would like to thank them for their consistent professionalism.

I am especially indebted to Alf Westwell for reading two drafts of the book and for the numerous comments which he made.

Finally, I would like to thank both my family for their support and my wife for her patience and encouragement.

I would gratefully like to acknowledge use of the following material:
Mason, J. *Learning and Doing Mathematics*, 2nd edn (York: QED, 1999), pp. 18–19, reproduced by permission of QED of York, qed@mathemagic.org
Pimm, D. *Speaking Mathematically* (London: Routledge, 1987), p. 20, reproduced by permission of Routledge.

Skemp, R. *The Psychology of Learning Mathematics* (Harmondsworth: Penguin, 1971), Copyright Richard R. Skemp, 1971, 1986, pp. 213–14, reproduced by permission of Penguin Books Ltd.

'Dust Disk Around a Massive Black Hole', Image supplied by AURA/ST ScI (STScI-PRC98-22 18 June 1998).

1 Setting the Scene

> This chapter aims to let you know:
>
> - a highly effective way of studying mathematics and its applications;
> - how well your current approach to study matches this ideal.

▶ What is mathematics?

Does it matter if you and your tutor have different attitudes to mathematics and its applications? You may believe that getting the correct answer to a problem is all that matters. For instance, when asked to solve the equation $2x + 3 = 11$ you might think that all you need to do is provide the answer $x = 4$. Unfortunately, your tutor is likely to have a different point of view. If the tutor is someone who applies mathematics he might also want a physical interpretation of the answer. And if he is a mathematician he is likely to be interested in how you reached the answer. Why is it the right answer? What has $x = 4$ got to do with $2x + 3 = 11$?

Clearly your attitude to mathematics and its applications has a significant impact on how well you succeed in your study. The diagnostic test in Figure 1.1 has therefore been designed to give you some insight into your own attitudes to mathematics. We will use your answers later in the chapter to see how your attitudes shape the way you study mathematics and its applications, so it will be worth completing the test now before you read on any further.

*Read the following six statements.
Then tick the three statements which best describe
your view of mathematics and its applications*

1. Mathematics consists of applying the rules of logic to solve problems involving numbers. ☐
2. Application of mathematics involves using and manipulating numbers to enable you to solve problems. ☐
3. Mathematics is the study of numbers, shapes and formulae representing numbers and shapes. ☐
4. Mathematics is a complex framework of ideas which enables you to make sense of the world. ☐
5. Mathematics is an abstract system of thought that is based upon a process of formal reasoning. ☐
6. Mathematics is concerned with solving problems about numbers and using the results to help you understand the world. ☐

FIGURE 1.1 SOME ATTITUDES TO MATHEMATICS AND ITS APPLICATIONS

▶ Facts or connections?

Many people view mathematics as a collection of facts, whether of rules, procedures, theorems, definitions, formulae or applications. Each fact can be clearly stated. We know, for instance, that $8 \times 7 = 56$ or that to add two fractions together you need to start by ensuring they share a common denominator. An application of mathematics then involves using these facts to solve practical problems, such as how to determine the optimal price at which to sell some goods. Ticking the boxes for statements 2, 3 and 6 in Figure 1.1 might suggest that this is how you view mathematics.

There is, however, a contrasting view that mathematics is more a system of interconnected ideas than a collection of facts. This view starts with the understanding that mathematics is underpinned by logical thought, so that we can guarantee that answers to mathematical problems are actually correct and can explain why they are. This view holds that different ideas within mathematics and its applications are inextricably linked with each other. Take the equation with which we started this chapter: $2x + 3 = 11$. This equation links together a whole range of different ideas. We have several numbers, the operations of multiplication and addition, the idea of a variable and the idea of equality all bound together in one statement. This view of mathematics is seen in statements 1, 4 and 5 in Figure 1.1.

Each of these two views of mathematics and its applications carries with it implications for your study. In order to tease out these implications, there is another diagnostic test in Figure 1.2 for you to complete. Again, please complete the test before reading any further.

*Read the following six statements.
Then tick the three statements which best describe your approach to studying mathematics and its applications*

1. Mathematics involves learning lots of formulae and rules. ☐
2. I try to make sense of a new application of mathematics ☐ by making sure I understand all of the underlying principles.
3. I learn mathematics by memorising all of the important ☐ results and by practising lots of problems.
4. Reading my notes several times over is my preferred way ☐ of studying mathematical applications.
5. The important thing is to see how well you understand ☐ an area of mathematics by tackling some difficult questions.
6. If I am introduced to a new concept, I look to see how it ☐ relates to other mathematical ideas.

FIGURE 1.2 APPROACHES TO STUDYING MATHEMATICS AND ITS APPLICATIONS

One approach to studying mathematics and its applications is to try to memorise all of the relevant information and to practise carrying out the tasks you are asked to complete. Memorising that $3 \times 1 = 3$, $3 \times 2 = 6$, $3 \times 3 = 9$ and so on is a classic example of this approach. Statements 1, 3 and 4 in Figure 1.2 have been designed to identify this approach to study. If you picked out most or all of these three statements, then is it also true that the previous diagnostic test indicated that you tend to view mathematics and its applications as a collection of facts? Theory suggests that if you view mathematics as a collection of pieces of knowledge, then it will seem quite reasonable to simply memorise each piece in turn.

A second approach to studying mathematics and its applications is to concentrate on making sure you understand what is going on. For this approach, memorising how to solve an equation is less important than understanding why the solution is valid and how the solution relates to other mathematical ideas. So you do not just use the letter 'x', you also make sure you understand what it represents. This approach to study is picked out in statements 2, 5 and 6 in Figure 1.2. Is it the case that if you view mathematics as a coherent system of ideas then you also tend to look for the meaning of ideas?

Most students will tend to approach their study either on the basis that mathematics is a collection of facts which need to be memorised or on the basis that mathematics is a system of ideas which need to be connected to each other. The two contrasting approaches are summarised in Figure 1.3. But which of these two contrasting approaches is more likely to be successful?

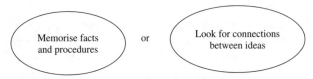

FIGURE 1.3 TWO APPROACHES TO STUDYING MATHEMATICS AND ITS APPLICATIONS

One of the most robust of all findings from educational research is that students who just try to memorise facts achieve lower grades than those who primarily seek meaning. If you think about it, the finding is entirely unsurprising. Facts are only relevant if they actually mean something to us, otherwise we quickly forget them.

It is also the case that some aspects of mathematics and its applications are not amenable to memorisation. For instance, Newton noticed a link between an apple falling and the mathematical idea that we now refer to as a vector. There is no formula you can memorise that will enable you to understand how gravity can be modelled by the idea of a vector.

Furthermore, approaching mathematics as a collection of facts encourages you to think that you can understand each fact on its own. Unfortunately for this approach, most mathematical ideas are neither quite so simple nor so isolated. Even the idea of an equation involves several mathematical ideas, as we have seen. In practice, you only appreciate most ideas when their links with other ideas are apparent. This leads us to two of the most fundamental of all intellectual skills: the skills of analysis (taking something apart) and synthesis (joining things together). Many of the skills we will consider in this book are in fact strategies for understanding mathematical ideas and completing mathematical tasks bit by bit, rather than all at once.

How you approach the study of mathematics and its applications will make a big difference to how well you succeed. Concentrating on memorising facts and procedures is not a recipe for success because it does not take account of the nature of mathematics and its applications. This book is designed to help you approach your study in a way that does take account of the nature of mathematics. And if you implement the advice given in this book, there is no reason why you should not find studying mathematics and its applications to be a rewarding activity.

> Effective study of mathematics is characterised by looking for connections between ideas rather than just memorising facts and procedures.

> **Reflection**
> Do you primarily look for meaning in your study or do you just tend to memorise facts and procedures?

Part I
Skills

2 Using Examples

> This chapter aims to:
>
> - demonstrate the value of using examples when studying mathematics and its applications;
> - enable you to use examples in order to understand ideas.

▶ Why use examples?

How do you teach a young child what a 'number' is? I doubt you would start with a theoretical discourse on the process of matching up sets of objects. You would start with actual examples of numbers. You might point to two cups and say 'two cups' or point to two chairs and say 'two chairs' and so on. Once the child had understood some specific numbers, you might talk about a 'number' as the word used to speak about any or all of the specific numbers.

Similarly, we can convey the more sophisticated idea of a 'real number' by considering a range of examples of real numbers. For instance, the following are all examples of real numbers: -6, π, 188, 2.43, -0.177, $\sqrt{2}$. From this we can see that real numbers include numbers which can be expressed as a ratio of two integers such as -6, 188, 2.43, -0.177 (for instance, $2.43 = 243/100$ and $-6 = -6/1$). Such numbers are called *rational* numbers. Real numbers also include numbers such as π or $\sqrt{2}$ which cannot be expressed as a ratio of two integers and are thus called *irrational* numbers.

It is important to note here that we need to consider a varied collection or set of examples. It would be of little use to include only integers in our set of examples of real numbers. To do this would give us a misleadingly incomplete picture of what a real number is.

In general, a mathematical idea can be conveyed by considering a varied collection of examples of the idea. Indeed, one of the clearest indications of whether you understand a mathematical idea is whether you can point out a variety of examples of the idea. Any new example of the idea is now simply 'one of those'.

Examples are also an important means of understanding applications of mathematical ideas. But before we take a look at the insight gained from considering a set of examples of an application of a mathematical idea, it is worth being clear about exactly what an application of a mathematical idea is.

We can see in Example 2.1 that the number of steps between two objects in some sense represents the real-world concept of distance. We have *applied* the mathematical idea of counting to the problem of determining how far apart two objects are from each other.

> **EXAMPLE 2.1** MODELLING THE CONCEPT OF DISTANCE
>
> Stand by a plant at one end of a garden and pick out some second plant at the other end of the garden. How far are you away from the second plant? How can we make sense of this question in a mathematical fashion?
>
> Perhaps the simplest way to do this is to walk towards the plant while counting the number of steps that you take. The *number* you count up to is then said to represent the *distance* between the two objects, measured in terms of steps. Specific distances are now matched by specific examples of numbers. For instance, you may have been looking at a plant that was 20 steps away from the plant by which you were originally standing.

> We say that the mathematical idea of a number *models* or represents the real-world concept of the distance between two objects.

It is important to note that mathematical ideas model our conception of the real world, rather than the real world itself. In the above example, we are not modelling two plants, we are modelling the idea that these two plants are some distance apart from each other. We can summarise our understanding of what an application is in Figure 2.1.

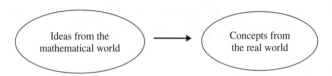

FIGURE 2.1 MATHEMATICAL IDEAS CAN REPRESENT CONCEPTS FROM THE REAL WORLD

So, for any application of a single mathematical idea we will have the mathematical idea itself and the particular real-world concept which the mathematical idea represents.

> Thus, to understand an application of a mathematical idea, we need both a set of examples for the mathematical idea we are applying and a matching set of real-world examples.

In the case of Example 2.1, we would take specific distances between different plants in the garden as examples of the real-world concept of distance. Each of these specific distances is then matched by a real number, as we can see in Figure 2.2. Clearly, some distances are short and some distances are long. Correspondingly, some real numbers are small and some real numbers are large. So the varied set of examples that we see in Figure 2.2 gives us a feeling for our mathematical model of distance.

However, providing a genuinely varied set of examples of an idea is not always as easy as it sounds. Sometimes it is difficult to come

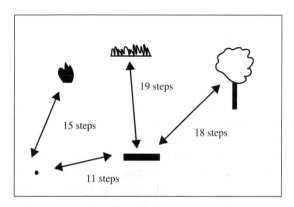

FIGURE 2.2 A SET OF EXAMPLES OF THE REAL-WORLD CONCEPT OF DISTANCE AND A MATCHING SET OF REAL NUMBERS

up with even one example of an idea. In the rest of this chapter we therefore explore some techniques for creating a genuinely varied set of examples.

> - Creating a varied set of examples of an idea helps you to understand the idea.
> - Applications of mathematics involve understanding both concepts from the real world and the mathematical ideas that model them.

Exercise 2a

1. Which mathematical idea is used to model the concept of time?
2. Which mathematical ideas are used to represent the real-world concept of area?

▶ Creating examples

Perhaps the easiest way to try to create a new example is to adapt an existing one. Take the idea of a function, which is after all one of the most important ideas in mathematics. If we allow x to be any real number, then we can informally say that the following rule provides us with an example of a function:

$$h(x) = x^2 + 3$$

If we adapt this example, we can speculate that the following rules also give examples of functions:

$$g(x) = x^2 - 3, \quad f(x) = x^2 \pm 3$$

It is, however, surprisingly easy to start with an example of an idea, make one small change, and end up with something that is not an example of the idea at all. Clearly, we need some way of telling whether or not we have actually created a genuine example. To deal with this issue, mathematicians come to agreements on the exact nature of the ideas they deal with. For each idea, this agreement is recorded in a short statement called a *definition* which spells out exactly what the idea is.

> Definitions ensure common agreement about what an idea actually is.

So we will now check whether or not we have just created two new examples of functions. In particular, we will check whether we created examples of a specific kind of function, a real-valued function. Now a common (but simplified) definition of a real-valued function is given as follows:

A *real-valued function* is a rule which assigns to each real number a single real number.

We can compare the two examples that we tried to create above with each element of this definition.

1. In both of our potential examples we can apply the rule to each real number and the rule always assigns a real number. For instance, for g the real number 3.25 is assigned to the real number 2.5 (since $(2.5)^2 - 3 = 3.25$). Or for f the real numbers 9.25 and 3.25 are both assigned to 2.5.
2. However, only in one of these cases – the case of g – does the rule assign to each real number a single real number. The rule $f(x) = x^2 \pm 3$ always assigns two distinct real numbers to each initial real number, so f is not an example of a function. We can see this contrast between f and g in Figure 2.3.

> In summary, we have taken each aspect of the definition in turn and looked to see whether the potential example meets all of the conditions. We can call this *analysing* a definition.

Given the complexity of most definitions this analysis is essential, otherwise you may fail to check an essential part of the definition.

If the suggested example does not meet the definition of the idea, then we have what we can call a *non-example* of the idea. Non-examples are particularly useful when they are almost, but not quite, an example. It is, after all, little use to say that the number two is not a real-valued function. That is obvious. It is, however, much more helpful to know that $f(x) = x^2 \pm 3$ is a non-example of a real-valued function.

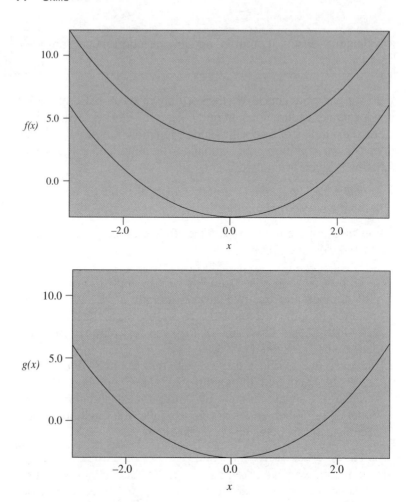

FIGURE 2.3 GRAPHS FOR THE RULES f AND g

Non-examples help to clarify our understanding of what qualifies as an example.

When looking for examples of mathematical applications, we will similarly need to check whether we have a genuine example or not. But in addition to checking our potential example against the definition of the relevant mathematical ideas, we will also need to check the definitions of the relevant concepts from the real world. We can see this in Example 2.2.

> **EXAMPLE 2.2** THE CONCEPT OF THE SPEED OF A BODY
>
> The speed of a body is the distance travelled by the body per unit of time. This is a definition of the concept of speed. So how do we model this concept mathematically? We use the mathematical idea of a real number to model the distance travelled per unit of time.
>
> We can now try to create both a real-world example of the concept of speed and a matching mathematical example. Let us take a plane flying the distance between London and Paris in one hour. This distance travelled by the plane in one hour is an example of the concept of the speed of a body. We can say this because in this case each element of our definition of the concept of speed is present. We have a body travelling a certain distance per unit of time.
>
> > In order to have an example of a concept from the real world, every part of the definition of the concept needs to be met.
>
> Furthermore, say the distance between London and Paris is 200 miles. So the distance travelled per unit of time is modelled by the real number 200. This number is the mathematical example that matches our real-world example of the concept of speed.

Simple examples

Once you have been able to create at least one new example of the idea it is worth creating some *simple* examples. These kinds of examples are usually relatively easy to understand, and for that reason alone they are very useful. So what makes an example simple? This question is perhaps best answered by looking at some simple examples.

The real-valued function given by $f(x) = x$ is certainly simple. This is the function that associates every real number with itself. By contrast, we can agree that the real-valued function $f(x) = x^6 - (2x - 3)^3$ is more complicated. After all, it takes some work to find out which real numbers are assigned to any given real numbers.

And what would constitute a simple real-world example of the speed of a body? We could take a jet fighter flying a distance of 1 mile in one hour. The mathematical example that matches this real-world example is of course the real number 1. But while this mathematical example is relatively simple, the real-world situation itself is certainly unusual. So for a simple example from the real world, I would be more likely to take a car travelling a sensible distance per hour.

> The important point in creating a simple example is to look for one that is easy to deal with.

Typical examples

The next stage in creating a range of examples is to find some *typical* examples.

> We can think of a typical example as one that displays all the characteristics of the idea. No simplifications are involved.

For instance, you could argue that the real-valued function given by $f(x) = 3x - 1$ is a typical example of a real-valued function. This at least avoids associating every real number with itself, as in the case of our simple example above. And for a typical example of the concept of speed, I would choose a person walking a distance of 3.1 miles per hour.

Unusual examples

Our final stage in creating a range of examples is to find some *unusual* examples. Again, the easiest way to see what constitutes an unusual example is to look at some of them.

There is, for instance, plenty of scope for unusual real-valued functions. One of the most unusual I have come across is the real-valued function f which assigns to every rational number the number 1 and to every irrational number the number 0. More concisely we can define this function by the following rule:

$$f(x) = \begin{cases} 1 & \text{if } x \text{ is rational} \\ 0 & \text{if } x \text{ is irrational} \end{cases}$$

As far as unusual examples of real-world concepts are concerned, each element involved in the concept can be unusual. Think, for instance, of unusual bodies. What about a tarantula travelling towards your toe a certain distance every second, or take the distance travelled by a shaft of light per year? How tedious to stick to a car travelling a sensible distance in a comfortable unit of time!

> Unusual examples test your understanding of an idea, to see if you really have understood what is going on.

Exercise 2b

1. Find a collection of examples for the following ideas. You should include the first example you can think of, two simple examples, two typical examples and an unusual example. Test each of your examples against a definition of the idea to make sure that you have genuinely found an example. In addition find a non-example for each idea.
   (a) A set.
   (b) A vector.
   (c) The tangent of an angle α.
2. Provide a set of examples for each of the following concepts from the real world. Further specify which mathematical idea models the concept from the real world and provide examples of the mathematical idea to match each of your examples of the real-world concept.
   (a) An incline.
   (b) The tension in a string.
   (c) The quantity of some merchandise demanded.

▶ Particular instances of examples

As well as looking at examples, it is often possible to look at just part or a single instance of an example. This allows us to focus on just one aspect of the idea with which we are concerned.

> We can call such a single instance of an example a *particular instance* of an example.

Take the example of a real-valued function given by the rule:

$$f(x) = x^4$$

This function assigns to each real number its fourth power. So we can focus on a specific real number – a particular instance – and see which real number is assigned to it by the function. Consider the particular instance of 2.5. If we apply the rule, we see that $f(2.5) = (2.5)^4 = 39.0625$. So we can see that the function assigns to the number 2.5 the number 39.0625.

Of course, some examples are too simple to yield any particular instances of themselves. The number 6 is an example of a number, but it does not give rise to any particular instances of itself: it is too simple. But by contrast the function $f(x) = x^4$ gives rise to countless particular instances. We can take other particular instances than just 2.5. What about 2, 0, π, -6.1, or any other real numbers you care to choose? Which real numbers are assigned to these four real numbers? Clearly, 16 is assigned to 2, 0 to 0, π^4 to π and 1384.5841 to -6.1.

There are plenty of ways of creating particular instances of examples. One common way is to replace a letter with an appropriate number. For instance, if you are dealing with a function that sends the real number x to the real number x^2 then check to see what happens to a specific real number, as we did above. In general, if an example involves a set, like the set of all real numbers **R**, then a particular instance would be a single member of the set. We can see this in the following identity:

$$x(x + 2) = x^2 + 2x \text{ for all } x \in \mathbf{R}$$

Note that $x \in \mathbf{R}$ is the notation used to indicate that x is a member of the set of all real numbers. So this equation holds for all members of the set **R**, including the particular instance of the number 7. Here we can see that $7(7 + 2)$ and $(7^2 + 2 \times 7)$ are equal to each other, both being 63.

Furthermore, we can look for typical, simple and unusual particular instances of examples. For our real-valued function given by $f(x) = x^4$, the number 2.5 could count as a typical instance, while 0 and 1 could be classed as simple. But you could also say that 0 and 1 are unusual instances because both of these numbers are sent to themselves by the function. No other numbers are sent to themselves by this function, so that certainly makes them unusual.

> Just as we created a varied set of examples for an idea, it is often helpful to create a varied set of particular instances for an example.

Once you have created a range of particular instances of an example it is also worth extracting the maximum benefit from your work. Do you notice anything about all of the numbers to which the real numbers 2, 0, π, −6.1 are assigned? All the numbers assigned to them are positive. Insights like this help you to make sense of an example, but such insights are usually only had − especially with

> **EXAMPLE 2.3** MODELLING THE SPEED OF A CAR
>
> Consider plotting the speed of a car (measured in miles per hour) against time (measured in hours), as given in Figure 2.4. The graph in Figure 2.4 defines a real-valued function, which we can call $v(t)$. The function v assigns to each instant of time t a speed $v(t)$. Here we have modelled the real-world concept of speed over a period of time using the mathematical idea of a function.
>
> We can now consider some particular instances of this application. We can take the particular instances of the starting time of the journey, when the car is not moving (and which has $t = 0$, $v(0) = 0$ as the corresponding mathematical particular instance), the time at which the speed is greatest (we can see from the graph that this first occurs when $t = 5$ and $v(5) = 80$), and the time at the end of the journey (which occurs when $t = 30$ and $v(30) = 0$).
>
>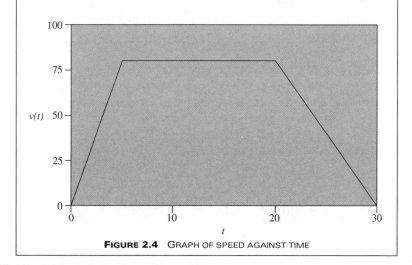
>
> **FIGURE 2.4** GRAPH OF SPEED AGAINST TIME

more complicated examples – when you take the trouble to create enough particular instances.

We can also consider particular instances of examples in the context of applications of mathematics. Parts of mathematical examples can model aspects of the real world, as we can see in Example 2.3.

Clearly, in this above example we can devise particular instances of the real-world examples and match them with particular instances of the mathematical examples.

> A particular instance enables you to focus on a single aspect of an example.

Exercise 2c

1. Create a range of particular instances for the following examples. Include the first particular instance that comes into your head, a typical case, two simple cases and an unusual case.
   (a) The set of all irrational numbers.
   (b) The real-valued function given by $f(x) = x^7 - 7$.
2. Create a collection of particular instances for the following real-world example. Further, point out the mathematical particular instances that match these particular instances.

   Size of a population over a given period of time as specified by the following graph:

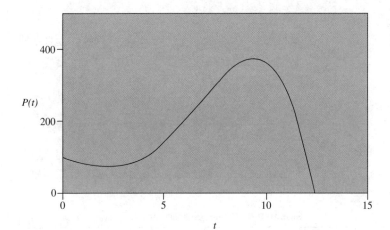

3. Which collection of examples below more effectively conveys the mathematical idea of a set? Why?
   (a) $\{2,5,9\}, \{3,7,11\}, \{2,4,5,8\}$.
   (b) $\{3, 0, -11.5\}, \{5\}, \{\pi, a, 21\}, \{1,2,3,4,\ldots\}, \{\}$.
4. Choose an idea that you are struggling to understand. Devise a varied set of examples for the idea. If appropriate, pick one of your examples and devise a varied set of particular instances for the example.

Extension Exercise 2

1. Devise a set of examples for each of the following mathematical ideas.
   (a) An abelian group.
   (b) A linearly independent set of vectors.
   (c) An injective function.
2. Devise a set of real-world examples for each of the following concepts. Each real-world example should be accompanied by the mathematical example that models it.
   (a) The centre of gravity of a body.
   (b) The rate at which a population decays over a given period of time.

> **Reflection**
> Have you ever tried to learn a mathematical idea simply by memorising its definition? How successful was your learning?

Summary

- Creating a varied set of examples of an idea helps you to understand the idea.
- A good set of examples will include typical examples, unusual examples and simple examples, and it will be complemented by some non-examples.
- Identifying particular instances enables you to focus on different aspects of an example.

3 Thinking Visually

> This chapter aims to:
>
> - demonstrate the value of thinking visually when studying mathematics and its applications;
> - help you to draw visual images and put them to good use.

▶ Why use visual images?

The discovery of DNA was one of the greatest scientific advances of all time. Watson and Crick realised there was a connection between our genes and a mathematical shape, a helix with two chains (as illustrated in Figure 3.1). They used an abstract mathematical shape to model actual shapes in the world around us, and their model enabled them to explain the relevant scientific data. We see here an instance of the way in which visual thinking plays a key role in applications of mathematics.

Visual thinking also plays a variety of other roles in mathematics and its applications. For instance, a visual image – whether a graph, a

Figure 3.1 A double helix

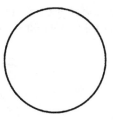

FIGURE 3.2 A CIRCLE

diagram, or so on – may give an overview of a mathematical idea or problem. Compare Figure 3.2 with the following definition of a circle:

> a circle is the set of all coplanar points that are all an equal distance from a specified point.

Clearly, the picture of a circle gives a better overall impression of what a circle is than the definition is able to. Another role that images perform is to provide insight into the different parts of an idea, and how these parts are related to each other. It is hard to tell from the definition of a circle how all the points in the set are related to each other, but we can clearly see in the picture of a circle how one point lies very close to some points and far away from others.

The rest of this chapter is designed to improve your ability to think visually. If we return to our description of the discovery of DNA, we can note that there were three elements to the visual thinking involved: creating a model; interacting with the model; relating the model to data about our genes. We will concentrate on a version of these three elements: drawing an image, looking at the image, and linking the image to other ideas.

- Visual images can model the real world, provide an overview of an idea, and provide insights into the different parts of an idea.
- There are three main stages to visual thinking in studying mathematics: drawing an image, looking at the image, and linking the image to other mathematical ideas.

Exercise 3a

1. Describe to a friend what a trapezium is (you are only allowed to use words)!

2. List some situations in the real world that can be modelled by a branching process, as for instance illustrated below:

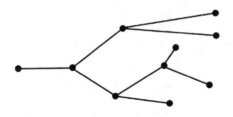

▶ Drawing visual images

We will consider two strategies for drawing visual images: the first strategy is to start by concentrating on the overall *structure*; the second is to begin by drawing the *details* of the image. These two strategies are closely linked to the roles we have seen that visual images play in mathematics of providing an overview and giving insight into the different parts of an idea.

Starting with the overall structure

Many visual images have an easily identifiable structure. Simple geometric shapes clearly fall into this category, but so do many diagrams for applications of mathematics. For instance, a problem involving a swinging pendulum is easy to picture. The first strategy, then, is to start by drawing the overall structure of the image.

> Once the structure is complete, you can fill in the details.

Take the problem from applied mathematics given in Example 3.1.

> **EXAMPLE 3.1** A LADDER LEANING AGAINST A WALL
>
> It is a classic problem to analyse the forces that act on a ladder that is propped up against a wall. This situation is easy to visualise straight away, so we can start by drawing the overall structure of the diagram. After the structure is in place we can then go on to fill in all of the details, whether the length of the ladder, or the forces acting at different points of the ladder.

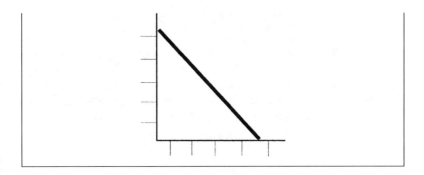

Starting with the details

You will, however, often find that you have to draw complex images for which you cannot quickly perceive the overall structure. For instance, try instantly to picture the graph of the real-valued function given by the following:

$$f(x) = \frac{(x+2)}{(3x-1)e^x}$$

The way to proceed in this kind of complex situation is to use the second of our two strategies, as we see in Example 3.2.

Start by drawing in individual details of the image, until the overall shape begins to emerge.

> **EXAMPLE 3.2** SKETCHING THE GRAPH OF A FUNCTION
>
> The aim is to sketch the graph of the real-valued function given by
>
> $$f(x) = \frac{2}{(x+1)}$$
>
> We start with the details of the graph that are easiest to fill in, which will be a selection of points. We know that $f(0) = 2$, $f(1) = 1$, $f(2) = 2/3$, $f(-2) = -2$, $f(-3) = -1$, and so on. We can plot a selection of points on the graph, but what happens in between the points we have plotted? After all, this particular function applies to all but one of the real numbers, not just a selection of them. In this case there are other details we can find out: how the slope of the graph behaves, where large real

> **EXAMPLE 3.2** *Continued*
>
> numbers are mapped to, why x is not allowed to be -1 and what happens for values of x near this unusual point. You can find explanations of how to determine these details in a standard textbook. Armed with all of these details we can begin to draw the graph of the function. We can see that a rough picture of the graph emerges naturally from all of these details.

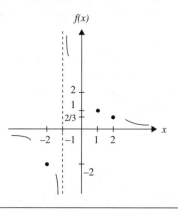

It is finally worth saying that you may need to adjust your expectations in order to apply this second strategy effectively. If you have always been used to dealing with images that you can instantly picture, then you will need to recognise that complex images demand a different approach. Do not worry if you cannot see the overall structure straight away; just go ahead and begin with the details.

> It may be helpful to start by concentrating on either the *structure* or the *details* when drawing a visual image.

Exercise 3b

Which of the above two strategies might you apply when drawing a visual image in each of the following cases, and why?

1. The graph of the real-valued function $f(x) = \sin x$.
2. A problem involving a weight sliding down an incline.
3. The tangent to the curve $y = \log_e \left(\dfrac{x^2 + 1}{2 - x} \right)$ at $x = 0$.

▶ How to use visual images?

Seeing images flexibly

Pablo Picasso is reported to have said that 'Observation is the most significant element of my life, but not any kind of observation.' It might have mattered how Picasso observed images, but does it matter how mathematicians see them? How can you ensure that the way you look at a visual image will be of use in your study of mathematics and its applications? Well, take a look at Figure 3.3. What do you see?

Do you see a cat or a collection of different geometrical shapes? If you focus on the overall impression, then you obviously see a cat. But if you focus on the details you can see several triangles, a parallelogram and a square. What is more, you can *choose* whether you see the cat or the geometrical shapes.

> Just as we could concentrate on the overall impression or the details when drawing an image, it is also possible to either look at the image as a whole or to look at the details.

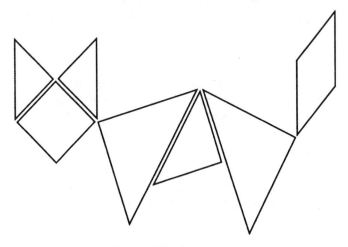

Figure 3.3 A tangram

In Example 3.3 we see how this applies to the graph of a function.

EXAMPLE 3.3 VIEWING THE WHOLE OF AN IMAGE AND VIEWING THE DETAILS

We can give an overall impression of the graph of the function $f(x) = e^x$ and pick out some details from the graph. The overall impression of the graph below is that of a steeply increasing curve. As for some details, we can see that when x is a large negative number the value of e^x is close to zero and also that the curve crosses the axis labelled $f(x)$ when $x = 0$ and $f(x) = 1$.

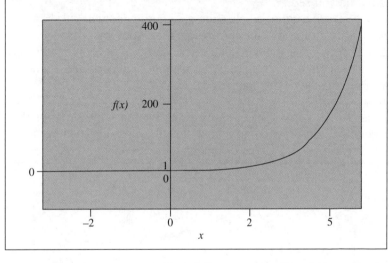

In practice, however, we often fail to exercise this choice of looking at the overall impression or focusing on the details. We simply see whatever our previous habits have conditioned us to see. But looking at images in different ways can allow you to exploit new avenues of thought, as we shall see in the next section.

> When you look at a visual image take a step back: alternate between concentrating on the overall impression and focusing on the details.

Linking visual images to the rest of mathematics and its applications

Visual images do not inhabit the world of mathematics on their own. Mathematics and its applications also involves writing, notation, models, proof, definitions, and so on. We have opened up a new approach in this chapter: either to concentrate on the overall impression or to focus on the details. Now is the time to use both of these approaches to throw light onto the rest of your study.

We can now try to link both an overall impression of a visual image and some details from the image to related ideas. To see this we can return to the function that we considered in the last example (see Example 3.4).

> **EXAMPLE 3.4** CONNECTING AN IMAGE WITH FORMAL MATHEMATICAL IDEAS
>
> Consider linking our earlier overall impression of the graph of the real-valued function given by $f(x) = e^x$ with a related mathematical idea. We pointed out earlier that the curve increased steeply as x increased. More formally we can observe that the gradient of the curve corresponding to any point x is given by $f'(x) = e^x$. And of course the larger the value for x in this equation then the steeper the curve will be; that is to say, that as x increases the curve becomes progressively steeper.
>
> As far as the details are concerned, we have already noticed that as x becomes a larger negative number then e^x gets closer to zero. Again more formally we can say that the limit as x tends to $-\infty$ is 0. And furthermore, we can note that the function f assigns to 0 the real number 1. This is of course where the curve crosses the axis labelled $f(x)$.

When you are dealing with an image that concerns an application of mathematics, it is important not only to relate the image to formal mathematical ideas but also to relate it to the real world. We can see this in Example 3.5.

> **EXAMPLE 3.5** CONNECTING AN IMAGE WITH THE REAL WORLD
>
> Consider a mathematical model of an incline, as given in the image below. Now my overall impression of this image is of a wedge shape that is wider on its left hand side and narrower towards its right. In the real world this represents a constant downward slope. It would of course also be possible to link these ideas with the mathematical idea of the gradient of a line.

30 Skills

> **EXAMPLE 3.5** *Continued*
>
> As for the details, we can see that the highest point of the incline is 10 m higher than its base, and the horizontal distance between the beginning and the end of the incline is 20 m. We could of course connect these ideas to more formal mathematics such as tangents and gradients.
>
>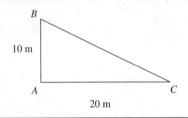

> Make sure that you link your images both with formal mathematical ideas and with the real world.

Exercise 3c

1. Can you link the following definition of the integral of the function given by $f(x) = x^2$ between 0 and 2 to the graph of the function?

    The *definite integral* of f, between 0 and 2, is the area of the region on the Cartesian graph which is bounded by the graph of the function, the lines $x = 0$ and $x = 2$ and the x-axis.

2. Consider the following graph of the acceleration, a, of a body as a function of time, t. Link both your overall impression and some

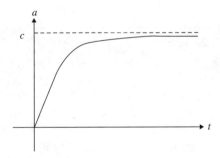

details from the graph to some formal mathematics and also to a real-world interpretation.

Extension Exercise 3

1. Draw a visual image to model the idea of a plane. Add some details to your image. Connect your work to some formal mathematical ideas.
2. Consider the image below, which concerns the gravitational force exerted by one body upon another body. Link this image to some formal mathematical ideas and to some further real-world concepts.

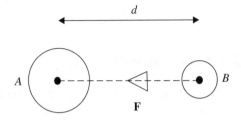

3. Link the following image to some formal mathematics:

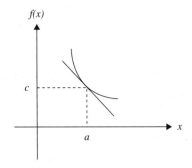

Reflection

Summarise, in your own words, the strategies covered in this chapter for drawing images, looking at images, and putting images to good use.

4 Coping with Symbols

> This chapter aims to:
>
> ensure that you can handle symbols intelligently and fluently.

▶ The purpose of symbols

Symbols abound in mathematics. You only have to pick up a textbook to see that this is so, from the most basic of symbols – such as 1, 2 and 3 – to symbols such as $\int$ for an integral and $\lim_{n \to \infty} a_n$ for a limit. Applications of mathematics also rely on symbols, with letters such as E, m and c used to represent aspects of the real world in Einstein's equation:

$$E = mc^2$$

Sadly, many people lose interest in mathematics when symbols start to proliferate. As soon as the mysterious 'x' is introduced, half the population is left behind; and letters such as 'x' are only the start of it.

So why are symbols relied on so much in mathematics if they cause so much difficulty? It is after all possible to write mathematics using only English. For instance, instead of writing $E = mc^2$ we could write the following:

> The energy of an object is equal to the mass of the object multiplied by the square of the speed of light.

To appreciate why symbols play such an important role in mathematics we need to clarify exactly what mathematical symbols are.

A symbol represents a mathematical idea: for instance, take the symbol π. This stands for the irrational number which represents the ratio of the circumference of a circle to its diameter. Or take the symbol 1043, which represents the number one thousand and forty three.

There is a difference between a symbol and the idea which we can say is represented by the symbol. Take a road sign that indicates the way to London, as in Figure 4.1. You would not try to reach London by driving up the road sign itself! It is not the road sign itself that is important: it is the information it conveys that matters. The same is true in mathematics; it is the idea which the symbol conveys that matters.

FIGURE 4.1 A ROAD SIGN

> You need to focus on understanding what each symbol actually means.

However, not only do symbols represent mathematical ideas, they do so in a *compact* fashion. This is one of the main reasons why symbols play such an important role in mathematics. It is obviously easier to write out Einstein's famous equation in symbols than it is in words. Or again, the symbol

$$\frac{dy}{dx}$$

is an excellent example of the conciseness of mathematical notation. This symbol stands for the instantaneous rate of change of some variable y with respect to a related variable x.

Such abbreviation also allows us to calculate much more easily. Imagine trying to solve the equation $4x + 3 = 2x + 6$ in the following manner:

> Four times the real number designated by the letter x plus three is equal to two times x plus six. This is equivalent to saying that four times x is equal to two times x plus three, which is again equivalent to saying that two times x is equal to three. This is the same as saying that x is equal to one and a half.

How much easier it is to write out the following solution:

$$4x + 3 = 2x + 6$$
$$\Leftrightarrow \quad 4x = 2x + 3$$
$$\Leftrightarrow \quad 2x = 3$$
$$\Leftrightarrow \quad x = 3/2$$

These two roles of representing ideas and providing a convenient shorthand are in tension with each other. One role focuses on interpretation, the other on ease of use. The aim of the rest of this chapter is to allow you to manage this tension.

Symbols are containers for meaning and aids to fluent calculation.

Exercise 4a

1. Think of a symbol or sign from everyday life. What purpose does it serve?
2. Take the following assemblage of symbols:

$$\theta \frac{\sqcup \sqrt{-}}{\Sigma \geq} \diamond \; \nearrow_{t_\odot}^T$$

   (a) Why are you unable to make much sense of this arrangement of symbols?
   (b) What impression do these symbols make on you?

▶ Finding meaning in symbols

We will first of all explore some ways of ensuring that you are fully aware of the actual meaning of the symbols you use. It is possible to think of symbols as taking on meaning at two distinct levels. The first level of meaning is the somewhat superficial level of the words used to refer to the symbol. For instance, we would refer to the symbol π as 'pi' or to the symbols

$$\int_0^1 x^2 dx$$

as referring to 'the integral of x^2 between zero and one'.

However, it is the second level of meaning that is more interesting. This is the precise mathematical meaning of the symbol. At this level of meaning, π refers to the number 3.1415926... rather than the word 'pi'. And our integral more specifically refers to the area on a Cartesian grid bounded by the graph of the real-valued function x^2, the x-axis and the lines $x = 0$ and $x = 1$.

It is worth noting that mathematical ideas themselves are always precise. So mathematical symbols represent precise ideas. The symbol 1043 does not mean 'some large number' and π does not mean 'a number close to 3'. Hence the definition of a symbol is important. It is essential to find out exactly which idea any symbol represents. Half-remembered versions of the definition are not good enough.

> Learning to pay attention to definitions is a key to success in mathematics.

However, just because you are familiar with a symbol does not mean that you fully understand what the symbol actually means. You may not be using symbols with sufficient precision. We can see this in Example 4.1.

> **EXAMPLE 4.1** THE MEANING OF THE SYMBOL '='
>
> The symbol '=' was first used by the Englishman, Robert Recorde. Only his explanation can do justice to the invention:
>
> > I will sette as I doe often in woorke use, a paire of paralleles, or Gemowe (double) lines of one lengthe, thus: ====, bicause noe 2. thynges, can be moare equalle.
>
> The sign '=' means that one mathematical object is equal to another mathematical object. So we are allowed to write

> **EXAMPLE 4.1** *Continued*
>
> $2 \times 2 = 4$ but not $2 \times 2 = 4 \div 2 = 2$, since 2 is certainly not equal to 4. You must not use the sign '=' to mean 'and then you get'.

You also need to be aware that the meaning of a symbol may depend on a variety of circumstances. For instance, the group of symbols (2, 3) could mean several different things depending on the context. We could be talking about a point on a Cartesian grid, or the set of all numbers between 2 and 3. Position, orientation, size, repetition, order and even the type-face of the symbol may all be relevant. There are also a large number of different conventions in operation. One textbook follows one convention, a different textbook will follow another. So make sure that you pay attention to the context.

In applications of mathematics, it is further important to ask how the symbol relates to the real world. What does the symbol mean in the particular real-world context being considered? The symbol m, for instance, often refers to the mass of a particular object, while the symbol Π may refer to the size of a given population.

> - Make sure that you know the precise meaning of every symbol that you use, in the context in which you are using it.
> - Relate symbols used in applications of mathematics to the real world.

Exercise 4b

1. Write down *from memory* a full and precise definition for each of the following symbols or groups of symbols:
   (a) $A \subset B$
   (b) $\equiv$
   (c) $\Rightarrow$
   Now find a formal definition for each of these in a textbook. How do your definitions compare with the 'official' definitions?
2. Which of the following usages of the symbol **R**, which stands for 'the set of all real numbers', are valid?
   (a) 2 is an **R**.
   (b) π is member of **R**.
   (c) The complex number $2 + 3i$ can be divided into **R** and imaginary parts.

▶ Developing fluency with symbols

Once you have grasped the meaning of individual symbols, the next stage is to make sure that you can fluently use the symbols together. Mathematics is in many ways a language, so it should not come as a surprise that we make use of an exercise that is usually associated with learning a foreign language to help us do this.

Translation

Our strategy is to translate the full meaning of mathematical texts into English prose. You need to be able to do this without referring back to your notes or to a textbook. And note that the focus here is on what the symbols mean rather than on how we pronounce them. We can see how to translate a text in Example 4.2.

EXAMPLE 4.2 TRANSLATION OF A TEXT

The text

$$\int_1^2 (2x^2 + 1)dx = \left[\frac{2}{3}x^3 + x\right]_1^2$$

may be translated as 'the area of the region on a Cartesian grid bounded by the curve defined by y equals twice x squared plus one, the x-axis and the lines defined by x equals one and by x equals two is equal to the real number determined by subtracting two-thirds of the cube of one added to one from two-thirds of the cube of two added to two'.

When we are looking at an application of mathematics, we not only need to spell out the mathematical meaning of the text, we also need to be aware that many symbols will stand for examples of concepts from the real world. So when translating a text related to an application of mathematics we need to include specific reference to these concepts. We can see this in Example 4.3.

EXAMPLE 4.3 TRANSLATION OF ANOTHER TEXT

The equation **F** = m**a** may be read as 'f equals m a' and translated as 'the vector denoted by **F** (which represents the force acting on

> **EXAMPLE 4.3** *Continued*
>
> a given body) is equal to the real number denoted by m (which represents the mass of the body) multiplied by the vector denoted by **a** (which represents the acceleration of the body)'.

In practice you will find that a text might well include some symbols you are very familiar with and other symbols which you do not fully understand. In this case you will find it useful to translate the less familiar symbols but leave unchanged within the text the symbols you know more thoroughly. We can see this, for instance, if we partially translate the text used in Example 4.2 above (see Example 4.4).

> **EXAMPLE 4.4** PARTIAL TRANSLATION OF A TEXT
>
> The text:
>
> $$\int_1^2 (2x^2 + 1)dx = \left[\frac{2}{3}x^3 + x\right]_1^2$$
>
> may be translated as 'the area of the region on a Cartesian grid bounded by the curve defined by $y = 2x^2 + 1$, the x-axis and the lines $x = 1$ and $x = 2$ is equal to ($\frac{2}{3}$ 2^3 + 2) − ($\frac{2}{3}$ 1^3 + 1)'.

It is only through repeatedly translating texts that you will be able to develop the kind of appreciation for the meaning of symbols that is needed for the more advanced study of mathematics and its applications.

Using symbols in calculations

Once you can easily translate the meaning of a text into English prose you are ready to leave behind the actual meaning of the symbols for a time. And as we saw when solving the equation $4x + 3 = 2x + 6$ earlier in this chapter, mathematics often does involve carrying out calculations without always thinking about the meaning of the symbols. As David Pimm (1987, p. 20) notes, 'Much of the computational fluency which mathematicians achieve with symbolic manipulation arises as a result of being able to work solely on the symbols themselves without thinking about their meanings.'

Even so, it is not advisable to entirely leave behind meaning even when manipulating symbols. Pimm (1987, p. 20) makes the following point: 'The crucially important skill, however, is the ability to re-integrate the symbols with their meanings *at will*, in order to interpret or check the details or results of such symbolic calculations.'

For instance, when solving our algebraic equation $4x + 3 = 2x + 6$, while it makes sense to solve this by manipulating symbols, it is important to remember that the equation is only satisfied when the real number designated by the letter x equals 3/2. And in this case we can see that it is indeed true that

$$4 \times \frac{3}{2} + 3 = 2 \times \frac{3}{2} + 6$$

Checking the meaning of the symbols you are using also provides a final means of developing a fluent use of a set of symbols. It helps in maintaining the balance between focusing on meaning and carrying out calculations.

Develop a fluent use of a set of new symbols by

- Writing out the full and precise meaning of the text
- Checking the meaning of symbols during calculations

Exercise 4c

1. Solve the inequality:

$$2x - 3 < x$$

   Further, spell out the full meaning of both the original inequality and the final stage of your solution.

2. Provide a translation of the following text:

$$\mathbf{F} = m \frac{d\mathbf{v}}{dt}$$

   where m is the mass of a particle, $\mathbf{v}$ is its velocity and the particle is acted upon by a force $\mathbf{F}$.

3. Choose a text of your own. Write out its full meaning.

Extension Exercise 4

1. Write out the full meaning of the following texts:
   (a) Let A and B be sets. Then $A \subseteq B \Leftrightarrow A \cap B = A$.
   (b) $\mathbf{a} \times \mathbf{b} = ab \sin \theta \hat{\mathbf{n}}$.
2. Solve the following differential equation:

$$\frac{dy}{dt} = y$$

Spell out the full meaning of both the original equation and the final stage of your solution. Further, given that y represents the size of a population at the time t, provide a physical interpretation of both the original equation and the final stage of your solution.

Reflection

- Are you in the habit of working with symbols that you do not fully understand?
- If so, what concrete steps will you take to break this habit?

Summary

- Symbols are containers for meaning and aids to fluent calculation.
- Make sure that you know the precise meaning of every symbol that you use.
- Develop a fluent use of a set of new symbols by writing out the full and precise meaning of the text and checking the meanings of symbols during calculations.

5 Taking Ideas Apart

This chapter aims to:

- convince you of the value of taking ideas apart when studying mathematics and its applications;
- enable you to take apart ideas in order to understand them.

▶ Why take ideas apart?

Some ideas are easy to understand. We know from direct experience what 'a book' is. We have seen enough books to understand what the idea of a book is in general. Unfortunately not all ideas are like this. Take the idea of a 'black hole'. You may have seen plenty of images connected with black holes, such as that from the Hubble Telescope given in Figure 5.1. But it is unlikely that you fully understand what a black hole is from these images.

A *definition* of a black hole might help. A black hole is a field of such strong gravitational pull that matter and energy cannot escape from it. However, unless we understand all of the simpler ideas employed in this definition we are none the wiser. What is a field? What is gravitational pull?

There is a similarly wide range of mathematical ideas. You only need to see a few triangles to realise what 'a triangle' is in general. But take the idea of an algebraic expression. We can consider some examples of algebraic expressions, such as $3(x + 6) - 5.35$, $2y^2 - x$ and so on. These examples involve such ideas as a real number, addition, multiplication and a variable. In fact, most of the ideas we deal with in mathematics and its applications depend on a host of other ideas, and often these ideas can depend on still other ideas; mathematics is a *hierarchical* subject.

What if you do not understand the idea of a variable? How then can you fully understand what an algebraic expression actually is?

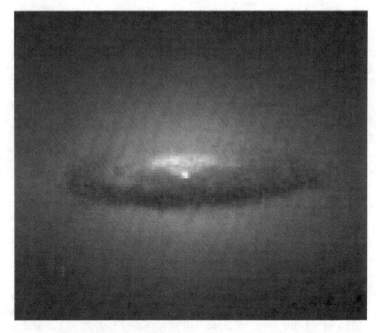

FIGURE 5.1 DUST DISK AROUND A MASSIVE BLACK HOLE

For any readers who are unsure, we will see later in this chapter exactly what a variable is. Suffice it to say for now that a variable is not 'a number that varies' or simply 'an unknown number that you need to find out'.

So if you are trying to learn an advanced mathematical idea and you have failed to understand even one of the basic ideas on which it depends, you are in trouble. You may never gain full understanding of the new idea and, what is more, mathematical ideas are easy to forget. It is not sufficient to have seen these basic ideas a few months or years ago and only have a vague understanding of them now. The understanding needs to be fresh and thorough.

> To learn a new idea you need to understand *all* of the more basic ideas on which the new idea depends.

It is therefore worth seeing which basic ideas contribute to the new idea. We can call this 'taking an idea apart' or, more formally, 'analysing a concept'. Once you have taken a new idea apart – and found out what the basic ideas are that contribute to it – you can then make sure that you understand each of the basic ideas.

> Take the new idea apart in order to make sure you understand all of these basic ideas.

Exercise 5a

1. What prevents you from understanding the idea of a 'language' when it is defined as follows: 'Let Σ be an alphabet. A subset S of the set of all words over Σ is called a language over Σ'?
2. State the standard trigonometric identity for $\cos x - \cos y$, without reference to any text.

▶ How to take an idea apart

In some ways it is easy to take an idea apart. You simply list all of the more basic ideas that help to make up the idea. This first of all requires finding an explanation of the idea, usually its *definition*. You then simply work your way through the definition or explanation of the idea, noting all of the basic ideas that are involved. See Example 5.1.

EXAMPLE 5.1 TAKE APART THE IDEA OF AN ALGEBRAIC EQUATION

We need an explanation of what an algebraic equation is.

- An algebraic equation is a statement that two algebraic expressions are equal to each other.

Working our way through this definition, we can list the following basic ideas: statement; the number two; algebraic expression; equality. It also helps to visualise our analysis, which we can do as follows:

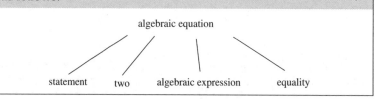

Once we have listed the basic ideas on which an idea depends, it may be possible to take the whole process a stage further.

> We can see that an idea depends on a list of basic ideas. On what even more basic ideas does each of these ideas depend?

Advanced ideas usually depend on several layers of more basic ideas. Note that you need to be able to state the definition or give an explanation of each of the basic ideas in order to move on to the next layer. If we return to the above example, we need to consider whether each of the ideas involved depends on any more basic ideas (see Example 5.2).

EXAMPLE 5.2 TAKE APART THE IDEAS OF A STATEMENT, THE NUMBER TWO, AN ALGEBRAIC EXPRESSION AND EQUALITY

We can work our way through each of these ideas.

- A statement – this idea does not seem to depend on any other ideas at all. But ideas are not always as simple as they first appear, as we shall see in Chapter 6.
- The number two – other than for philosophers of mathematics, ideas do not come much simpler, so we can safely leave out any analysis of this idea.
- An algebraic expression – this is a collection of numbers (both variables and constants) linked together with the various operations of arithmetic. Working our way through this definition, we can list the following basic ideas: constants; variables; the operations of arithmetic. We can again visualise this analysis as follows:

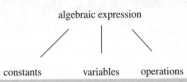

Equality – this is a statement that two potentially different mathematical objects actually have the same numerical value.

All of these ideas that contribute to the idea of an algebraic equation provide examples of the main types of mathematical idea. The idea of a 'statement' comes from the realm of *logic*. An 'algebraic expression' and 'the number two' are straightforward examples of mathematical *objects*. In this chapter we will concentrate on ideas that are objects, such as numbers, triangles and functions. The operations of arithmetic, which contribute to the idea of an algebraic expression, are called *processes*. For instance, we can think of the operation of addition as the process of adding two numbers together to form a new number. Finally there is the idea of 'equality'. This is an idea which expresses a *relationship* between two mathematical objects. In this case we can say that two algebraic expressions are related to each by being equal to each other.

> The main kinds of mathematical ideas are: ideas from logic; objects; processes; and relationships.

In conclusion, we can summarise the basic strategy of taking an idea apart as follows:

> **How to take an idea apart**
>
> 1. State the definition or give an explanation of the idea.
> 2. List all the basic ideas involved.
> 3. If appropriate, repeat the above two steps for each of the basic ideas involved.

This strategy can also be used both to take apart concepts from the real world and to take apart applications of mathematics. For instance, if you want to understand the concept in chemistry of the *rate of a reaction* one way to help you do this is to analyse its definition:

The rate of a reaction is the derivative of the concentration of the reactant with respect to time.

Clearly in order to understand this concept from chemistry you need to understand the ideas of a derivative, reactant, concentration of a reactant and time. It usually takes more work to analyse an application of mathematics, because our list of more basic ideas will need to involve both ideas from mathematics and concepts from the real world. We can see this in Example 5.3.

> **EXAMPLE 5.3** ANALYSIS OF NEWTON'S SECOND LAW OF MOTION
>
> Here we consider a mathematical model of the relationship between several concepts from the real world. Newton proposed that the resultant external force on a body, the mass of the body, and the acceleration of the body were all related to each other. He used the mathematical idea that we now refer to as a vector (a quantity that possesses magnitude and direction) to represent both force, **F**, and acceleration, **a**, and the idea of a positive real number to represent mass, m. He then proposed that these ideas were related to each other through the equation:
>
> $$\mathbf{F} = m\mathbf{a}$$
>
> So in order to understand this application of mathematics we need to understand the following concepts from the real world: resultant external force; body; mass; acceleration. And there are also the following mathematical ideas that we need to understand as well: vector; variable; real number; equality.

Exercise 5b

1. List the basic ideas on which each of the following ideas in italics depends:
   (a) The *momentum* of a particle is the vector quantity which is the product of the particle's mass with its velocity.
   (b) A *natural logarithm* of the positive real number a is the power c in the equation $a = e^c$. We write $c = \log_e a$.
2. (a) List the basic ideas on which the following idea in italics depends:
   The *tangent function* is the function which assigns to each real number x – where the real number x may be taken to represent an angle of size x radians – the single real number $\tan x$.

(b) The tangent of the acute angle x radians – $\tan x$ – is the fraction a/b, where a and b are the positive real numbers given in the following right-angled triangle:

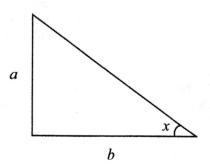

The basic ideas on which the real number $\tan x$ depends include the following: acute angle; fraction; variable; real number; and triangle. Which basic ideas have been omitted from this list?

▶ Include all of the basic ideas

Unfortunately, it is not always quite so easy to take an idea apart as the above section might suggest. For a start, it is relatively easy to overlook some of the basic ideas on which a new idea depends. What is more, it is likely that the basic ideas you overlook are the very ones that need revising.

- Some basic ideas are easy to overlook when making a list.
- These are usually the ideas that you most need to revise!

To illustrate these points, I would like you to analyse the following definition of a real-valued function:

A *real-valued function*, $f: x \to f(x)$, is a rule which assigns to each real number, x, a single real number $f(x)$.

Once you have analysed the definition, compare your list of basic ideas with the list given in the footnote below.[1] Did you omit any ideas from the list?

---

1. Rule; assignment; real number; variable; a single real number.

One of the main reasons why you might fail to include a basic idea in the list is because you do not understand it. For instance, did you include in your list the idea of an *assignment* between objects? In this case each real number x is paired with the real number $f(x)$. If you list only the basic ideas that you understand then the whole process becomes pointless! The aim is to identify the ideas you do not fully understand.

Another reason why you might fail to include a basic idea in the list is because the basic idea is hidden behind a symbol. Did you include in your list of ideas for the analysis of a real-valued function the idea of a variable? After all, the word 'variable' was not included in the definition itself. However, if we look through the definition, we see that the symbols x and $f(x)$ appear. And of course, these symbols are examples of variables.

> It is important to use the skills developed in the chapter on coping with symbols to identify the ideas that symbols represent.

We will, finally, explore this idea of a variable at greater length. What is a variable? An explanation by Richard Skemp (1971, pp. 213–14) is hard to improve upon:

> Often in everyday life we refer to an object in a way which indicates what set it belongs to, but not which particular element of the set it is. For example, 'a car' is a shorter way of saying 'an unspecified element of {motor cars}'. 'A handkerchief' means 'an unspecified element of {handkerchiefs}'. We talk in this way for various reasons. One is when it does not matter which: if we want to blow our nose, any handkerchief will do. Another is when we want to make statements which are true for each and every element of the set, and do not depend on which element we refer to. Examples: a car must have a licence; a car must stop at red traffic lights; a car needs petrol and oil.
>
> A third reason for not specifying which element of a set we are referring to is when we do not know: for example my wallet has been stolen by a pickpocket. In cases of this last kind, we often wish to know.

In short, we can say the following:

> a variable is an unspecified member of a given set.

If the variable is an unspecified member of the set of all real numbers (i.e., if the variable is an unspecified real number), then we call it a *real variable*. This way of thinking about a variable goes beyond the idea of a variable as 'an unknown that we need to find out'. It is worth ensuring that you fully understand the ideas you have to work with.

Exercise 5c

1. A *relation R* on a set *S* is a statement which is true for some ordered pairs of elements from *S* and false for others. List the basic ideas on which the idea of a relation depends. Compare your list with the list given in the solutions to this exercise. If you omitted any idea or ideas, can you explain what the ideas are?
2. The *definite integral* of a real-valued function between the real numbers *a* and *b* is the area of the region on a Cartesian grid bounded by the graph of the function, the lines $x = a$ and $x = b$ and the *x*-axis. List the basic ideas on which this idea of a definite integral depends. Ask a fellow student to do the same. Compare your list with the list of your fellow student. Compare both lists with the list given in the solutions to this exercise. Can you explain any ideas that either you or your fellow student failed to include?
3. Which idea or ideas have been missed out from the list of basic ideas in the following analysis:

    The basic ideas which contribute to the equation

    $$3x^{3/2} + 2x = x^3$$

    are the following: real number; variable; division; equality; addition; and multiplication.

▶ Revising basic ideas

Let us suppose you have come across an idea you do not fully understand. You follow the advice of this chapter and take the idea apart. You find that you do not understand one or two of the basic ideas that make up the new idea. Unfortunately, this is not the end of the matter. You now of course have to ensure that you understand each of these ideas.

The next step is to write down the symbols or words that you do not understand. This is an important aspect of getting to grips with your difficulty. Writing down will force you to face your lack of understanding.

> You now need to state the definition, or else to give a clear explanation, of each of the basic ideas concerned.

You will of course need to look in your notes or in relevant textbooks. But if you cannot find an explanation of an idea, this is not an excuse for giving up. All you have to do is ask one of your fellow students or your tutor.

Exercise 5d

1. Find an explanation for each of the terms in the phrase 'the sine of the angle $\pi/4$ radians'.
2. Select several ideas that you do not understand and take each idea apart. Revise any of the basic ideas involved that you cannot confidently explain.

Extension Exercise 5

1. Take the following ideas apart and give an explanation of each idea that is involved.
   (a) A group G with respect to a binary operation $\circ$ is a non-empty set G on which the binary operation $\circ$ is defined and for which the following axioms hold:
   - closure – $a \circ b \in G$ for all $a, b \in G$;
   - associativity – $(a \circ b) \circ c = a \circ (b \circ c)$ for all $a, b, c \in G$;
   - identity – there exists $e \in G$ such that $a \circ e = a = e \circ a$ for all $a \in G$;
   - inverses – for each $a \in G$ there exists $a^{-1} \in G$ such that $a \circ a^{-1} = e = a^{-1} \circ a$ for all $a \in G$.

   (b) A real-valued function f is continuous at a point a in its domain if and only if for all $\varepsilon > 0$ there exists $\delta > 0$ such that
   $$|x - a| < \delta \Rightarrow |f(x) - f(a)| < \varepsilon$$

2. List the mathematical ideas and the real-world concepts which contribute to the following text.
   The concentration, c, of a single reactant is determined by the following equation:
   $$-\frac{dc}{dt} = k$$
   where k is a constant.
3. Find a definition of each of the following ideas, take each definition apart, and find an explanation for any of the ideas involved that you do not fully understand.
   (a) Simple Harmonic Motion.
   (b) An injective function.
   (c) A linearly independent set of vectors.

> **Reflection**
> Which exercise in this chapter did you find most helpful? What did you learn from the exercise?

Summary

- To take an idea apart state the definition, or give an explanation, of the idea and list all of the basic ideas involved.
- Taking an idea apart allows you to see whether you understand all the basic ideas on which the idea depends and to revise any basic idea you do not fully understand.

6 Thinking Logically

This chapter aims to:

- convince you of the importance of thinking logically when studying mathematics;
- enable you to understand the key elements of mathematical logic.

▶ Why think logically?

Imagine what life would be like if a suspect could be convicted of a crime simply on the *intuition* of a police officer. We would do away with long drawn-out trials. Unfortunately, different officers would have different intuitions. Miscarriages of justice would be the norm. Society would fall apart.

Can we rely on intuition in mathematics, even if we cannot rely on it in a court of law? Well, consider a semi-circle of diameter 2, as shown in Figure 6.1a. Clearly, the length of the arc for this semi-circle is π. Now in Figure 6.1b, each of the two smaller semi-circles has a diameter of length 1 and an arc of length $\pi/2$. So the total length of both these arcs is π. Then in Figure 6.1c we have four even smaller semi-circles, each of which has a diameter of length $1/2$ and an arc of length $\pi/4$. So the combined length of all of these arcs is again equal to π. We can continue to inscribe even smaller semi-circles in our diagram, but in each case the combined length of the arcs will equal π. As these semi-circles get smaller and smaller, their arcs will eventually just form a straight line, but the length of the line is 2. Hence $\pi = 2$.

Intuition has clearly led us astray! How can we have π equal to both 3.14159... and to 2? If mathematicians relied solely on intuition

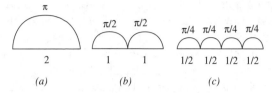

FIGURE 6.1 CALCULATING π

to guide their work then they would be led to believe contradictory results. The mathematical equivalent of miscarriages of justice would be the norm. Mathematics would fall apart.

Some historians speculate that in the sixth century BC the Greek mathematician Thales observed discrepancies in the mathematics of the time. Egyptian and Babylonian rules for calculating the area of a circle, for instance, were different. Since a circle has a unique area, at least one of these rules must have been incorrect. As a result, Thales may have seen the need for a strictly rational basis on which to establish mathematics.

> This rational approach that unifies mathematics is called 'logic'. It aims to rule out discrepancies in mathematics.

Logic is concerned with establishing truth on a sound footing. However, the way that truth is established in logic is completely different from the way that truth is established in everyday life, or in the sciences, for example. In everyday life we often establish truth in an *informal* fashion. One person might assert something, and we do not believe them. But if someone whose judgement we trust makes exactly the same point, we believe them. The context plays a huge role. Meanwhile, in the sciences the search for truth proceeds on the basis of making observations and collecting evidence. Theories are put forward to try and explain the evidence. We call this approach to establishing truth *inductive*. Other types of search for truth also follow this inductive pattern, including those suggested proofs of the existence of God which are based on the collection of evidence from experience.

By contrast, logic establishes truth in a manner that is both *formal* and *deductive*. Logic is first of all formal because everything is precisely defined. No allowance is made for the context. Instead, everything follows carefully agreed rules. Second, logic is deductive because truth is established on the basis of agreed starting points, called *axioms*. New results then follow from these axioms according to the carefully agreed rules. For instance, the geometry of Euclid starts from a set of axioms. Examples of these axioms are: 'It is possible to draw a unique straight line between any two points'; 'All right angles are equal'; and 'It is possible to draw a circle with any specified centre and radius.' Starting from these axioms, Euclid was able to prove that a whole host of geometrical relationships must be true. For instance, he could prove that the sum of the angles of a triangle always equals 180°. Whatever the values of x, y and z in Figure 6.2, we know that they add up to 180°.

It is, however, possible to vary any set of axioms, even Euclid's axioms: absolute truth is not to be found in mathematics.

Still, you should be able to see that a formal and deductive approach to establishing truth avoids many pitfalls. In everyday life words can be used with different meanings, depending on the context in which they are spoken. Just think about the questions 'Tea or coffee?' and 'Salt or pepper?' Clearly, the first question could receive any of the answers: 'Tea, please', 'Coffee, please', or 'Neither, thank you'. However, 'Tea and coffee, please' is not an acceptable option. By contrast, we would be quite happy to serve someone with both salt and pepper. We cannot allow any such ambiguity in logic. As we shall see later in this chapter, 'or' is chosen in logic to mean 'either one, or the other, or both'.

Inductive approaches can also lead to error. A new experiment might be conducted that confounds earlier theories. Einstein's theory of relativity was able to explain observations that Newtonian

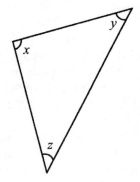

Figure 6.2 The angles of a triangle

mechanics could not account for; yet Newton's mathematics has stood the test of time. Consider an example of what can easily happen if we try to follow inductive approaches in mathematics (see Example 6.1).

> **EXAMPLE 6.1** INDUCTIVE REASONING IN MATHEMATICS
>
> Can we guarantee that $n^2 - n + 41$ is always a prime number if n is a positive integer? Try collecting evidence for this. If $n = 1$, then $n^2 - n + 41$ equals 41, which is a prime number. For $n = 2$, we get 43 which is also a prime number. Try a few other values and see if you can predict what happens. The outcome is discussed in the note below.[1]

If you try to apply the rules of everyday life, or of science, to establishing truth in mathematics then you will never learn to study mathematics effectively. The formal and deductive approach is central to mathematics. Even if you only intend to apply mathematics in science or engineering, you still need to understand some basic logic. It is worth spending some time learning the rules of the game.

> Mathematics is established on the basis of formal deductions rather than on intuition, informality, or experimentation.

Exercise 6a

1. Can we guarantee that if we draw lines between n points on the circumference of a circle then the circle will be divided into 2^{n-1} regions?

---

[1] All of the values $n = 3$, $n = 4$, $n = 5$, and many others, lead to prime numbers. But we only need to find one value of n which results in a non-prime number to spoil everything. Try $n = 41$. This leads to a number which is certainly not prime.

2. What is the number $0.\dot{9}$? Is it the number 1, or is it a number slightly less than 1?
3. How does the usage of the word 'and' differ in the following two sentences: 'I tried to cross the road and I got knocked over by a car'; 'It rained yesterday and I like walking'.

▶ An introduction to logic

Statements

The idea of a *statement* is one of the most fundamental ideas in logic. In everyday life, people make a variety of statements: 'You should have voted in the last election'; 'Évariste Galois died on 30 May 1832'; 'Eat your greens.' In the realm of logic, however, the idea of a statement is carefully defined.

> A sentence that is capable of being either true or false, but not both, is called a *statement*.

Now the sentence 'Évariste Galois died on 30 May 1832' is a true sentence. By contrast, the phrase 'Eat your greens' is not a sentence that is either true or false; it is simply a command.

In mathematics we make statements about mathematical objects, such as numbers or sets, but these statements must still be either true or false. For instance, the sentence '$3 \times 4 = 13$' is a statement. It is in fact false, but that does not make it anything less of a statement! By contrast, the expression '3×5' is not a statement. How can it be true or false? We need to claim something is true in order to create a statement.

There is one final nuance that is also worth getting across. We saw in the last chapter that we often wish to make claims about unspecified members of a given set: claims about variables. We could claim, for instance, that X is a triangle. As it stands, however, we cannot tell whether this claim is true or false. It all depends on what X is. The object X could be a square for all we know. But once we know what X is, it is easy to tell whether the claim is true or false.

> A sentence that is either true or false, but not both, when a value is substituted for a variable is also called a statement.

A classic example of this kind of statement is the linear equation. On the face of it, we cannot tell whether the statement '$2x + 3 = 5$, where x is some real number' is true or false. Solving this equation, however, lets us know that $x = 1$ is the only value of x which ensures that the statement is true.

Logic is, of course, not content with simply making statements that are (or can be) either true or false. The usual aim of a valid logical argument – which we call a *proof* – is to establish whether or not a given statement is actually true. We will look at proof in greater detail in Chapter 10. For now, we will first of all look at one of the most important type of argument employed in logic, that of implication.

Implication

An *implication* is the idea that if a given statement is true, then a further statement must also be true. We say that statement A implies statement B, if B is sure to be true when A is true. We can say that the statement 'T is an equilateral triangle' implies 'T is an isosceles triangle'. Clearly, if T is an equilateral triangle, then it is bound to be an isosceles triangle as well. Another way of looking at the idea of implication is to say that we deduce the truth of B from the truth of A. It is, however, worth noting that saying A implies B does not mean that A *causes* B, but simply that B will certainly be true when A is true.

The usual notation that expresses the idea of implication is the symbol '$\Rightarrow$': instead of saying 'A implies B' we write $A \Rightarrow B$. Clearly we can only use this symbol between two statements. We are certainly not allowed to write $2 \times 2 \Rightarrow 4$ because neither 2×2 nor 4 are statements.

Implication allows us to create a chain of deductions. We start with a statement A that we either know to be true or accept is true. The truth of statement A may then imply the truth of statement B. The truth of statement B may imply the truth of statement C, and so on until we establish that the final statement in the chain is true.

> We can think of a chain of deductions as a line of dominoes. If the first statement is true then we push the domino over, and all the rest of the dominoes fall.

We can see a chain of deductions in Example 6.2 below.

> **EXAMPLE 6.2** A DEDUCTION
>
> Consider the right-angled triangle below.
>
>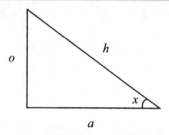
>
> Assuming Pythagoras' Theorem, the laws of arithmetic and the equation $\sin x = o/h$ and $\cos x = a/h$, we can deduce that $\sin^2 x + \cos^2 x = 1$ as follows:
>
> $$\begin{aligned} o^2 + a^2 &= h^2 \quad &\text{(Pythagoras' Theorem)} \\ \Rightarrow (o^2 + a^2)/h^2 &= 1 \quad &\text{(by the laws of arithmetic)} \\ \Rightarrow (o/h)^2 + (a/h)^2 &= 1 \quad &\text{(by the laws of arithmetic)} \\ \Rightarrow \sin^2 x + \cos^2 x &= 1 \quad &\text{(since } \sin x = o/h \text{ and } \cos x = a/h\text{)} \end{aligned}$$

However, we also say that some statement A implies a statement B when A is in fact false. In this case statement A gives no indication as to whether B is true or false. We have seen that 'T is an equilateral triangle' implies 'T is an isosceles triangle'. But what if it only turns out that T is not an equilateral triangle? T could still be an isosceles triangle; we simply do not know whether it is or is not.

The idea of implication is one of the most widely used ideas in mathematics, so it is no wonder that a whole range of words are used to convey exactly the same idea. Amongst others, we can use the words or phrases: hence; therefore; as a result; this implies; thus; this ensures that. But each word or phrase conveys exactly the same idea of implication.

> $A \Rightarrow B$ means:
>
> - if A is true then B is true;
> - if A is false then B either might be true or it might be false.

Exercise 6b

1. Which of the following sentences are statements?
   (a) $3 = 2$.
   (b) $\int_0^2 (x^2 + 3)\,dx$.
2. Decide whether the following statements are true or false.
   (a) $\int_0^1 x^2\,dx = 1/3$.
   (b) The number π can be written in the form p/q, where p is an integer and q is a positive integer.
3. Which of the following arguments are valid?
   (a) T is an isosceles triangle $\Rightarrow T$ is an equilateral triangle.
   (b) $x < 1 \Rightarrow x < 2$, where x is some real number.

▶ Further logic

The realm of logic includes far more than the rules concerning statements and implications. Given how important logic is to mathematics, it will be worth looking at some further ideas. In particular, we will consider ways in which complex statements can be formed from more basic statements and we will look at a more complex form of argument than implication.

Not

It is often the case that we need to say that a statement about some mathematical objects or processes is not true. For instance, we may need to assert that two objects x and y are not equal to each other. If a statement A is true then the *negation* of A is called NOT A. If A is true then NOT A is false, while if A is false then NOT A is true. Clearly, we know that the statement $3 = 2$ is not true. But we can say that the statement $3 \neq 2$ is true.

And

We saw in Exercise 6a that the word 'and' can take on different meanings in everyday life. In logic, however, the word 'and' is solely used to assert that two statements are both true. For instance, we can claim that '$2 \times 2 = 4$ and $2 < 3$' or that '$5 < 2$ and $2 + 7 = 9$'. Clearly, '$2 \times 2 = 4$ and $2 < 3$' is a true statement while '$5 < 2$ and $2 + 7 = 9$' is a false statement.

Or

We can also connect two statements together with the word 'or'. We use the word 'or' to mean that either one of the statements is true, or the other statement is true, or both statements are true. All that is ruled out is the possibility that both statements are false.

For instance, we say that '2 = 3 or 5 = 5' is a true statement. In this case, only one of the two contributory statements involved is true, but that is enough. We can also say that '2 = 3 or 5 = 6' is a false statement, and that '2 = 2 or 5 = 5' is a true statement.

Quantifiers

We saw earlier in this chapter that a sentence which is either true or false when a value is substituted for a variable is called a statement. It might, however, be the case that every possible such substitution creates a true statement. This might seem rather far-fetched, but in practice this situation crops up frequently. Take the following sentence: $2x + 2 = 2(x + 1)$, where x is a real number. Clearly, this equation is true whatever value x takes. So we can say that the following statement is true:

For all real numbers x, $2x + 2 = 2(x + 1)$.

However, it is not true to say that for all real numbers x, $2x + 2 = 0$. All you need to do is substitute the value $x = 5$, or plenty of other values, to see that the equation is not true for all real numbers x. But what we can say is that there is at least one value of x for which $2x + 2 = 0$ holds. Or more formally:

There exists at least one real number x, such that $2x + 2 = 0$.

We thus have sentences that are true for all the values of the variable, and sentences that are true for at least one value of the variable. In both cases we have 'quantified' how many values of the variable ensure that the sentence is true. For this reason, the phrases 'for all' and 'there exists' are called *quantifiers*.

Equivalence

We have seen that one way in which arguments are formed is through the use of implication. However, we will often find both that $A \Rightarrow B$ and that $B \Rightarrow A$. For instance, we know that if $x^2 = 1$ then $x = 1$ or $x = -1$, and that if $x = 1$ or $x = -1$ then $x^2 = 1$.

If two statements imply each other then the statements are said to be equivalent to each other, and we write $A \Leftrightarrow B$. So if two statements A and B are equivalent to each other, then A will be true when B is true, and B will be true when A is true. Again, it is worth noting that the sign $\Leftrightarrow$ is only allowed between two statements. You are certainly not allowed to say that $2 \times 2 \Leftrightarrow 4$.

What use is this idea of equivalent statements? Should the idea of implication not be sufficient to establish whether a statement is true or false? Well, look at an example (see Example 6.3).

EXAMPLE 6.3 IMPLICATION OR EQUIVALENCE?

Which values of the real variable x satisfy the following equation: $x - 1 = \sqrt{3x^2 + x + 1}$? We could validly argue as follows that

$$\begin{aligned} x - 1 &= \sqrt{3x^2 + x + 1} \\ \Rightarrow \quad x^2 - 2x + 1 &= 3x^2 + x + 1 \\ \Rightarrow \quad 2x^2 + 3x &= 0 \\ \Rightarrow \quad x(2x + 3) &= 0 \\ \Rightarrow \quad x = 0 \text{ or } x &= -3/2 \end{aligned}$$

Now you might be tempted to conclude from this argument that both the values 0 and $-3/2$ satisfy the original equation. However, if we interpret the symbol $\sqrt{}$ to mean the positive square root – as would usually be the case – and if we substitute either of the values 0 or $-3/2$ for x in the equation then we find that the equation is not satisfied. In fact, all that we can conclude from our argument is that if the first statement in our argument is true, then the final statement is true. But it so happens that the first statement is false.

In order to make any conclusions about which values of x satisfy an equation we need to employ an argument in which the equations are all equivalent to each other, as we shall see in greater detail in Chapter 10.

We will often need more sophisticated arguments than just the idea of implication.

Exercise 6c

1. Decide whether the following statements are true or false:
   (a) $2 \times 2 = 4$ or $\pi = 2$.
   (b) There exists $x \in \mathbf{R}$ such that $x^2 + x + 1 = 0$.
   (c) For all $x \in \mathbf{R}$ $3x = 2$ or there exists $x \in \mathbf{R}$ such that $x^2 > 5x$.
2. Which of the following arguments is valid?
   (a) $dy/dx = 2x \Leftrightarrow y = x^2$.
   (b) $x < 2 \Leftrightarrow x < 3$.

Extension Exercise 6

1. Can you spot any logical errors in the following arguments?
   (a) $\sqrt{x(2x-1)} = x - 2 \Leftrightarrow 2x^2 - x = x^2 - 4x + 4$
   $\Leftrightarrow x^2 + 3x - 4 = 0 \Leftrightarrow (x+4)(x-1) = 0 \Leftrightarrow x = -4$ or $x = 1$.
   (b) $\tan^2 x + 1 = \sec^2 x \Rightarrow (\sin^2 x / \cos^2 x) + 1 = 1/\cos^2 x$
   $\Rightarrow \sin^2 x + \cos^2 x = 1$. We know that this last statement is true. Hence the statement $\tan^2 x + 1 = \sec^2 x$ is true.
2. Why are the phrases 'if and only if', and 'necessary and sufficient' the same as 'equivalent'?

Reflection

- Describe – in some detail – how you went about studying this chapter on logic.
- How can you improve your study of the next chapter you tackle?

Summary

- Mathematics is established on the basis of logic rather than on intuition, informality, or experimentation.
- A sentence that is capable of being either true or false, but not both, is called a *statement*.
- An *implication* is the idea that if a given statement is true then a further statement is true, but if the given statement is false then the further statement could be true or false.

7 Making Connections

> This chapter aims to:
>
> give you insight into how to make connections between ideas.

▶ Connections in mathematics

Our brains consist of a mass of cells that are densely connected to each other. Synapses link brain cells to create an entity that escapes the imagination. Mathematics displays a similar complexity. Mathematical ideas are all just as densely connected to each other.

Connections in mathematics defy easy categorisation: for instance, mathematical ideas can unexpectedly represent each other. We find that the number 1 can be represented in terms of two irrational numbers e and π, and the complex number i (the square root of -1) through the equation:

$$e^{i\pi} = -1$$

Then mathematical ideas can be linked with each by being used together. At the simplest level we can link two numbers and the idea of addition by performing the operation of addition on the two numbers, or we can use ideas such as sets, members of sets, and rules in forming the idea of a function. Then there are logical relationships between ideas. We can prove that a given statement must be true, if a set of axioms are taken as given. We can create new mathematical ideas out of existing ideas. A function, after all, results from associating the members of one set with those of another. Finally, applications of mathematics link mathematical ideas with ideas from the real world: for example, we have real numbers linked to distance.

In summary, we can say that two ideas are connected to each other if they have something in common (if the ideas follow a similar pattern). Mathematics has been called the study of patterns. It could also be said that:

> Mathematics is the study of connections.

Why make connections?

Mathematicians who are the first to spot connections between seemingly disparate branches of mathematics are guaranteed enduring fame. Descartes gained part of his fame by linking algebra and geometry with each other. For instance, by mapping two-dimensional space on to a coordinate system Descartes enables us to represent a geometrical object, such as a line, by an algebraic equation.

So why are connections so important to mathematics? It is because virtually any mathematical task you attempt will make use of connections. We have in fact been making use of connections between mathematical ideas in all the earlier chapters. We have already linked examples with definitions, concepts in the real world with mathematical ideas, visual images with more formal mathematics, more complicated ideas with their constituent ideas, symbols with the ideas they represent and logical ideas with each other. Making connections is what the study of mathematics is all about.

Connections are essential elements of understanding. Early Greek mathematicians simply could not see how irrational numbers, such as $\sqrt{2}$, could exist alongside rational numbers. They were in fact so shocked that they tried to keep the existence of irrational numbers a secret. It is only as you are able to see how irrational numbers fit in with an understanding of rational numbers that you understand them. We, for instance, can appreciate the way in which both irrational and rational numbers can be expressed in decimal form, so that $\pi = 3.14159265\ldots$ and $1/3 = 0.33333333\ldots$, and so on.

Another way to see how essential connections are to mathematics is to consider the ineffectiveness of approaching the study of mathematics by memorising endless rules about what to do in any given situation. In the early stages of studying mathematics this may work reasonably well. You might, for instance, learn how to add two fractions together as set out in Example 7.1.

> **EXAMPLE 7.1** RULE TO ADD TWO FRACTIONS TOGETHER
>
> In order to add 3/4 to 5/8 we find the lowest common multiple of 4 and 8 – in this case 8 but in some cases a number that escapes us – and then somehow turn the bottom of each fraction into this number 8 so that we now have to add 6/8 and 5/8 to each other. We then add the tops of each fraction together to give 11 so the answer we are looking for must be 11/8.

Unfortunately, once you have learnt several hundred unconnected rules mathematics becomes a little harder to manage. Which rule are you supposed to apply in which situation?

> It is always easier to remember a set of ideas that are connected to each other in some way than to remember a set of completely disconnected ideas.

Clearly the more you try to ensure that you make connections between different mathematical ideas, the more effective you are likely to be in studying mathematics. The rest of this chapter aims to help you to do this.

> It is important to make connections when studying mathematics because:
>
> - understanding an idea involves connecting it to other ideas;
> - it ensures that you actually remember the ideas.

Exercise 7a

1. Spend two seconds trying to memorise the following list of words: eight, boat, quickly, why, rocking-chair, hence, way, deep. Now spend two seconds trying to memorise this list of words: two, six, ten, fourteen, eighteen, twenty-two, twenty-six, thirty. Which set of words is more memorable? Why?

▶ How to make connections?

Early attempts to make connections between mathematical ideas are likely to benefit from employing a *systematic* or structured approach. We have in fact already employed systematic approaches to finding connections in the earlier chapters. In this chapter, however, we focus directly on the process of how to make a connection in general.

What kind of connection should you look for?

The first step in our systematic approach to making connections is to be clear about the kind of connection you are looking for. If you take any set of mathematical ideas, there is likely to be a huge range of possible connections between the ideas. Some of these connections will be useful, but a large number of them will not. For instance, when you are trying to solve the equation

$$x^2 - 2x = 0$$

it will not be much use for you to know that the numbers 2, −2 and 0 are connected to each other through the equation

$$2 + (-2) = 0$$

It will, however, be much more useful to you to know that the expressions x^2 and $-2x$ are connected to each other because they each share the factor x. We are not just interested in any old connections; we are interested in connections that are *useful*.

In some ways, however, you can only build up an awareness of which connections actually matter with increasing experience of specific branches of mathematics. In algebra, for instance, you simply need to become aware of the fact that common factors are useful.

> You need to become familiar with the kinds of connection that matter in the particular branch of mathematics you are studying at the time.

In order to build up more experience of looking for connections we will first of all focus on two particular types of connection. The first connections on which we will focus are the connections between mathematical ideas and concepts from the real world that we encounter in applications of mathematics. We have already seen, for instance, in Chapter 2, that real numbers are linked to distance and that vectors are linked with gravitational force.

We will then look at connections that result from a process called *generalisation*. To understand what is meant by generalisation, take the following four equations:

$$61 = (30 \times 2) + 1$$
$$-15 = (-8 \times 2) + 1$$
$$211 = (105 \times 2) + 1$$
$$27 = (13 \times 2) + 1$$

We can see that these equations all have something in common with each other. Each equation states that a given odd number is equal to twice some specific integer, plus 1. This last sentence is a generalisation of the above four equations. Alternatively, we can say that each of the equations is a particular instance of the following statement:

Let a be any odd number, then $a = (2 \times m) + 1$ where m is an integer.

It is worth noting that generalisations frequently involve the introduction of a variable. Here we have introduced the variables a and m. As we saw in Chapter 5, variables allow us to make statements which are true for each and every element of a set.

> We can then say that each and every element of the set when substituted for the variable yields a particular instance.

Generalisation involves reversing the process carried out in Chapter 2, where we looked at particular instances of mathematics ideas. This time we generate a more general statement from a set of particular instances.

When are generalisations worth looking for? You will often find that you have the potential for a large – or even infinite – number of cases. The question then is, are they linked in any way? Are they in fact all particular instances of some more general idea, example or statement?

Spotting a connection

Once we know what we are looking for and when we should look for it, the next stage is to look systematically for the connection. In order to see whether two or more specific ideas are connected to each other, you need to *juxtapose* the ideas against each other (literally, put them next to each other). You cannot expect connections to emerge just by looking at one idea.

So, if you are trying to model a given real-world concept, juxtapose specific mathematical ideas against the concept you are trying to model. You need to see a genuine connection between the real-world idea and the mathematical idea. The number of steps you take to reach a plant at the other end of the garden is genuinely connected to how far away the plant is. Of course, some real-world ideas will be modelled more or less well by different mathematical ideas. It would be possible to model the distance between you and the plant by a set which contains the same number of members as the number of steps it takes to reach the plant. So if it took six steps to reach the plant we could model the given distance by a set containing six objects, like the set $\{a, b, c, d, e, f\}$, but this model is far less practical.

With generalisations you often need to put several cases next to each other to see if they are all particular instances of some more general idea, as we shall see in Example 7.2.

EXAMPLE 7.2   GENERALISATION OF A PAIRING OF NUMBERS

The table below indicates a link between each natural number a with another natural number b. Clearly, we can see that the numbers 5 and 2 differ by 3, that the numbers 8 and 5 differ by 3, that the numbers 11 and 8 differ by 3 and so on. So we can generalise to say that as the numbers in the top row increase by 1 then the numbers in the bottom row increase by 3. And we will see in the next section that this connection can be spelled out even more explicitly. It would, however, be difficult to spot this generalisation if you were only able to look at the first pairing of numbers.

a	1	2	3	4
b	2	5	8	11

> You will often find that you need to generate quite a few cases before you see any generalisation.

Spelling out a connection
It is one thing to put two or more ideas next to each other, and say yes, these ideas are obviously connected to each other; it is quite another thing to spell out exactly how the ideas are connected to each other. Mathematicians are not satisfied with two ideas being related to each other in some vague, undefined fashion. Any connection needs to be sufficiently clear that others can see it.

How then can we spell out a connection between a mathematical idea and a concept in the real world? The simplest way is to give them both the same name or label. For instance, take a body that is accelerating. We can both say that the acceleration of the body is denoted by **a** and give the vector

$$\begin{pmatrix} a \\ b \end{pmatrix}$$

the label **a**. This might all seem too simple to bother about, but in fact it is almost criminal not to indicate which mathematical idea is modelling which real-world concept!

> Give a concept from the real world and a mathematical idea the same label to indicate that they stand for the same entity.

It is similarly important to clearly spell out any generalisations that you make, as John Mason (1999, pp. 18–19) points out: 'Generalising is about making connections and capturing them in a succinct statement from which particular instances can be retrieved by specialising.'

Mason further calls the act of writing down what you have seen *crystallizing*. This is just as important as seeing a connection in the first place, because without a willingness to crystallise your insights you will not be able to communicate them effectively.

We can see this if we return to Example 7.2. We have already noticed a connection between the numbers on the top row of the table and the numbers on the bottom row, but so far our insights have not been very satisfactory. How can we spell out the connection between the two rows of the table more clearly? The solution is that each number on the bottom is three times the number on the top less one. More formally, we can say that the table associates n with $3n - 1$ where n is a natural number. So we have an example of a function, which we can call f and define by $f(n) = 3n - 1$ for $n \in \mathbf{N}$.

It is one thing, however, to claim that a set of cases can be seen as particular instances of some general statement; it is quite another thing to prove that a generalisation is actually true, as we saw in our earlier work on logic. It is one thing to notice that for every right-angled triangle you have ever seen, the square of the hypotenuse is equal to the square of each of the other two sides added together. This entitles you to say that

$$x^2 + y^2 = z^2$$

where x, y and z represent the lengths of the sides of the set S where $S = \{$every right-angled triangle I have ever seen$\}$. It is quite another thing to say that $S = \{$all right-angled triangles$\}$. Just as we need to test whether our model is in any sense a useful model, generalisations also need to be tested.

Employ a structured approach to looking for connections between ideas

- Ensure you are clear about the kind of connection you are looking for.
- Systematically juxtapose ideas to see if they are connected.
- Spell out the exact nature of any connection that you see.

Exercise 7b

1. Determine both the next few terms and the general term in the following sequence: 1, 7, 17, 31, 49, ...

2. Determine the relationship between force and distance as specified by the following table:

Distance	Force
3 cm	(1/9)N
2 mm	25 N
2.5 m	(1/62,500) N

3. Why is the following statement confusing?
Now $s = d/t$ so if we travel s miles at a speed of 20 miles per hour it will take $s/20$ hours.

▶ Connections in specific areas of mathematics

So far in this chapter we have focused on identifying two particular kinds of connection between ideas: the way in which a set of cases can be linked with each other by being particular instances of the same idea, and the way in which mathematical ideas model real-world concepts.

In many ways, however, the best way to appreciate connections in mathematics and its applications is to look at specific topics. To illustrate this we will now look at connections between ideas in two areas of mathematics.

Connections between trigonometric functions

The study of trigonometric functions – and in particular the functions sine, cosine and tangent – is an important area of mathematics, with applications in other fields. So what kinds of connection are evident between the ideas found here?

It is first worth pointing out that all the usual connections between ideas will be relevant here. You will want to link visual images with formal mathematical ideas. The members of a set of objects may all be particular instances of some general result. There will be a need to concentrate on the logical connections that are evident between the ideas in this field, and so on.

However, one of the most notable features of the trigonometric functions sine, cosine and tangent is that they are all defined on the basis of a right-angled triangle. The definition of each function depends in part upon the ratio of two of the sides of the triangle. So it should not come as a surprise that various mathematical results exist to connect these functions with each other. For instance, the real-valued functions given by $f(\theta) = \sin\theta$, $g(\theta) = \cos\theta$ and $h(\theta) = \tan\theta$ are linked to each other by the following result:

$$\tan\theta = \frac{\sin\theta}{\cos\theta}$$

Or again, the sine and cosine functions are connected as follows:

$$\sin a + \sin b = 2\sin\left(\frac{a+b}{2}\right)\cos\left(\frac{a-b}{2}\right)$$

If you are aware of these results it will then be possible to make use of them when solving specific problems. If you see a pattern emerging in a problem that is similar to the pattern of an established result then you might be able to take advantage of this but, as we have seen earlier in the chapter, you cannot just expect to see these patterns automatically.

> When you are working on a problem involving trigonometric functions, look through the established results to see whether one of them usefully connects ideas you are working with to each other.

It is finally worth noting that in most areas of mathematics and its applications you will need to look for results that connect the ideas you are working with. In particular, you need to learn to look through the central results in your field, and place them side by side against the material you are working with. Take some time to compare your work with each of the main results. Ideally you should aim to do this mentally rather than by actually glancing through the pages of your notes. Ask yourself whether there is any connection between these results and your work. Only then will you be able to spot where an established result links together ideas that you are working with.

> You cannot just expect connections between ideas to be become obvious to you of their own accord.

Extension material: connections to eigenvectors and eigenvalues

In many ways, eigenvectors and eigenvalues are sophisticated mathematical ideas, but it is also the case that they are closely related to one of the most fundamental of all mathematical ideas: the idea of a vector. It therefore helps to take advantage of this close connection between these ideas when trying to make sense of eigenvectors and eigenvalues.

Now eigenvectors and eigenvalues are typically defined as follows:

$$A\mathbf{x} = \lambda\mathbf{x}$$

where A is a square matrix, $\mathbf{x}$ is a vector (with an equal number of rows to the number of rows in the square matrix), which defines an eigenvector for A, and λ is a real number, which defines an eigenvalue for A. For instance, it is evident that

$$\begin{pmatrix} 2 & 0 \\ 0 & 2 \end{pmatrix} \begin{pmatrix} 3 \\ 1 \end{pmatrix} = 2 \begin{pmatrix} 3 \\ 1 \end{pmatrix}$$

and in this case we say that the vector

$$\begin{pmatrix} 3 \\ 1 \end{pmatrix}$$

is an eigenvector for the given matrix, with the value 2 as the associated eigenvalue for this eigenvector.

For many students this area of mathematics is simply about calculating eigenvectors and eigenvalues, without appreciating anything more. But how can we connect this information about eigenvectors and eigenvalues to the fundamental idea that underpins this area of mathematics: the idea of a vector?

We first of all need to spell out clearly what a vector actually is. A vector may be defined as a quantity that possesses magnitude and direction. We can now connect this understanding of a vector to the equation that defines an eigenvector and its associated eigenvalue. In particular, we can ask what happens to an eigenvector when it is multiplied by its matrix. And even more specifically, we can ask what happens to the magnitude and direction of an eigenvector when it is multiplied by its associated matrix.

The important point is to notice that the vectors $\mathbf{x}$ and $\lambda\mathbf{x}$ both have the same direction. All that happens to an eigenvector when it is multiplied by the associated matrix is that its magnitude changes; the resulting vector retains the same direction as the original vector. And when an eigenvector is multiplied by its associated matrix then its magnitude will be scaled by a factor which is called the eigenvalue.

When working with vectors it is often important to analyse the situation in terms of magnitude and direction. But more generally, you might want to take the definition of a central idea in an area of mathematics or one of its applications you are currently studying, and then try to make connections between this more fundamental idea and a concept you are struggling to understand or a problem you are struggling to solve.

After all, mathematics is characterised by fundamental ideas that occur again and again. So linking these more fundamental ideas to other ideas will both help you understand the ideas and give your work a sense of coherence.

> Juxtapose the definition of the fundamental idea against ideas you are working with. Only then are connections likely to become evident.

Extension Exercise 7

1. List some of the connections between ideas that are evident in the following areas:
   (a) Set theory.
   (b) Probability functions of a discrete random variable.
2. Take a look at a branch of mathematics you are currently studying. Take a look at a number of problems or theorems. Try to identify as many as possible of the useful connections between the ideas. Can you see any pattern emerging of the kinds of connection that are useful?

> **Reflection**
> Compare the approach to looking for connections that we have employed in this chapter with the technique used in Chapter 3 for linking visual images with more formal mathematics.

Summary

- You need to become familiar with the kinds of connection that matter in the particular branch of mathematics you are studying at the time.
- Employ a structured approach to looking for connections between ideas.

Part II
Tasks

8 Solving Problems

> This chapter aims to:
>
> improve your ability to solve problems in mathematics and its applications.

▶ Will any strategy do?

Many students try to solve problems by staring at the statement of the problem until they can see the solution. This kind of strategy works tolerably well with problems for which a standard method of solution exists. If you spot the relevant method then all you have to do is apply it, and if most of the problems you have met in the past simply involved using a standard method then it might seem reasonable to try to apply this strategy to all the problems that you face in future.

Unfortunately as your level of study gets more advanced you will find that many problems cannot be solved using standard methods of solution. You will increasingly be required to solve problems which at first you have no idea how to solve.

> The following problem, for instance, might possibly fall into this category:
>
> The first term of a geometric progression is 35. Given that the sum to infinity is not less than 40, find the least value of the common ratio.

There is, however, a more effective way of trying to solve complex problems than just to wait for inspiration. This is to employ a structured approach. There are several advantages to working in this way. First of all, a structured approach provides a framework in which to work without becoming discouraged. It also enables you to take greater control of your work as you can focus your attention more easily on the aspects of the problem that matter. And finally, even if you are not able to solve the problem, the experience gained of applying a structured approach will stand you in good stead for solving other problems.

> Some approaches to problem-solving are better than others.

The structured approach that we will follow here is based on work by George Polya. He divided problem-solving into four phases: understanding the problem; devising a plan; carrying out the plan; looking back over your work.

> We will illustrate this structured approach with the problem outlined above, our *illustration problem*.

▶ How to solve it?

1. Understand the problem
The first stage in our structured approach is to make sure that you actually understand the problem. It is surprising how many students try to solve problems without understanding the ideas involved. If you are dealing with a standard problem then you might be able to just mindlessly apply a stock procedure, but complex problems demand a different approach.

A particularly effective way of identifying what you need to understand is to answer the following two questions:

- What do I know?
- What do I want?

Answering these two questions helps you to focus your attention on the information given in the problem and on the goal of the problem.

> In the case of our earlier problem, we know that we have a geometric progression whose first term is 35 and that the sum to infinity must be not less than 40; we want to determine the least allowable value of the common ratio of the geometric progression.

> In order to fully understand the answers to these two questions you will need to employ skills that were covered in Part I.

To start with, you will need to *analyse* your two answers. List all of the ideas that are involved and provide a simple explanation of each idea. You will usually find it helpful to consider *examples* and *particular instances* of the ideas, and to rewrite what you know and what you want to know in mathematical *symbols*. You cannot usually expect to solve a problem when it is expressed in English prose, so you may need to *translate* the statement of the problem into the language of mathematics. As far as applications of mathematics are concerned, some of the symbols introduced will make explicit any *connections* between the real world and mathematical ideas. It may further be possible to represent the information *visually*. Make sure that you connect the visual image to the ideas in the problem. There may finally be *connections* between ideas in our initial explanation of the problem that can be pointed out.

> Now in order to solve our illustration problem, we need to understand what the following ideas are: a geometric progression; the first term of a geometric progression; the common ratio of a geometric progression. We shall take the idea of a least value as sufficiently obvious not to warrant further explanation. It is worth writing down a simple explanation for each of these other ideas, even if they are already familiar. So, a geometric progression is a sequence in which the ratio of any term to the previous term is always constant. The first term in a geometric progression is the first term in the sequence and the common ratio is given by the constant ratio of any term to the previous term.
> Before we look at any examples and particular instances it will be useful to introduce some notation. Fortunately in this case

> **Box** *Continued*
>
> there is some standard notation that we can draw upon. In general the terms of a geometric progression may be listed as follows:
>
> $$a, ar, ar^2, ar^3, ar^4, \ldots$$
>
> where a is the first term and r is the common ratio. To connect these symbols with the original statement of the problem, we can see that $a = 35$ and that the goal of the problem to find a least value of r. So if we took the example of the case $r = 2$ then we would have the geometric progression:
>
> $$35, 35 \times 2, 35 \times 4, 35 \times 8, \ldots$$
>
> Meanwhile, the case $r = 0$ gives the geometric progression:
>
> $$35, 35 \times 0, 35 \times 0, 35 \times 0, \ldots$$
>
> and the case $r = -1/2$ gives the example:
>
> $$35, 35 \times \left(-\frac{1}{2}\right), 35 \times \left(-\frac{1}{2}\right)^2, 35 \times \left(-\frac{1}{2}\right)^3, \ldots$$

2. Engage in planning

Once you have gained a genuine understanding of what you know and of what you want to know, the next stage is to begin to plan how to bridge the gap between them. This is of course the goal of all problem-solving. There are two strategies we will consider for bridging this gap between what is known and what is wanted. The first of these is to identify and draw on resources. The second is to ensure that you take control of your work.

It is often possible to link your problem to other relevant information, so you need to ask what additional resources you can draw upon to bridge the gap between what is known and what is wanted. Clearly to do this it is necessary to have gained a thorough grasp of the area in which you are working. Look through your notes or a textbook to see if any procedure or theorem is relevant. Does any mathematical result link what you know to what you want to know? You might follow the guidance given in Chapter 7 to help you make appropriate connections.

> So what resources can we draw upon to help us solve our illustration problem? Clearly it would be a great help if we were to be able to evaluate the sum to infinity of our progression. An inspection of the relevant chapter of an appropriate textbook (if the result is not already lodged somewhere in your mind) should be enough to find that the sum to infinity of a geometric progression is always given as follows:
>
> $$S_\infty = \frac{a}{1-r}$$
>
> provided that $-1 < r < 1$, and for values of r outside this range then the sum is either infinite or non-existent. A look back at our earlier example in which $r = 2$ confirms that the terms in the progression just keep on increasing so that we might expect such progressions not to have a finite sum. To link this new resource more clearly with our problem, we can say that the sum to infinity of our geometric progression is
>
> $$S_\infty = \frac{35}{1-r}$$
>
> provided $-1 < r < 1$. So ensuring the sum to infinity is not less than 40 is the same as ensuring the expression $35/(1-r)$ is not less than 40.

Once you have identified some of the relevant resources that you can draw upon you are now in a position to take control of your work by devising a plan. The aim is to propose a way of using the resources you have identified to bridge the gap between what you know and what you want to know.

Now in any given problem there will be a vast range of ways to try to bridge this gap. Some options will reach a solution quickly, others will reach a solution after plenty of work and yet others will never yield the solution. It therefore makes little sense to spend most of your time trying to carry out the first option that comes into your head as it could well fail to lead to a solution.

> Before you rush into a particular course of action, brainstorm as to which strategies might work by generating at least two different options.

> We now need to devise some plans for our illustration problem. The goal of the problem is to determine the least value of the common ratio of the geometric progression. Using our resources we know that we must ensure that $35/(1-r)$ is not less than 40. One plan might be to substitute different values of r into the expression $35/(1-r)$ in the hope that we can then pick out the least allowable value of r. Another plan would be to note that the phrase '$35/(1-r)$ not less than 40' suggests an inequality, so why not try to solve the inequality and see what happens?

It is worth noting that your options need only concern how to start work on solving the problem. There is no need to be convinced that any option will lead to a solution. The gap between what is known and what is wanted is often too wide to bridge straight away, even using additional resources.

3. Carry out a plan

Once you have devised at least a couple of plans, you need to carry out the option that you think is most likely to be effective. Make sure that you write down your work, as this often triggers new ideas about how to proceed. It is also important to monitor how well the solution is proceeding.

> You should feel tension between a certain measure of confidence that your plan might well lead to the solution, and a recognition that you could be wasting your time.

Is it worth continuing with this line of attack? Is the solution getting too complicated? As a general rule, the longer the solution is taking without any hopeful signs, the more likely you are to be heading towards a dead-end.

> In our problem, you might think that a lucky guess would let us finish things quickly, so why not start by trying this line of attack? If we try our earlier examples first of all, we can observe that if $r = 2$ then the sum to infinity will not exist because r is outside the range $-1 < r < 1$ so we can rule out this value of r. When

> $r = 0$ then the sum to infinity is $35/(1 - 0)$ which is unfortunately less than 40. And then when $r = -1/2$ the sum to infinity is $35/(3/2)$ which is again less than 40. My monitoring indicates that this plan does not seem to be leading us to a solution, and of course how would we know if we had found the least value of r anyway? Perhaps a better guess would yield a smaller value of r that still worked.

Now, in general, either your work will lead to a solution or your monitoring will suggest that you are heading towards a dead-end. Congratulations are in order if you manage to solve the problem with your first plan (although it will be worth checking your solution). But if the first plan fails then we simply move onto the next stage in the structured approach.

4. Review the situation

If a solution does not appear in sight then there are several options open to you. Note that giving up straight away is not on the list of options!

1. Review your work and check for mistakes. Making mistakes is an occupational hazard of being a mathematician, but leaving them uncorrected is another matter.
2. Try your second plan of attack on the problem.
3. Return to the original statement of the problem and repeat the stages of the structured approach. Make sure that you fully understand the problem. Are there any other resources you can draw upon to bridge the gap between what is known and what is wanted? Are there any other options for solving the problem?
4. Analyse how you have been tackling the problem. Perhaps the easiest way to do this is to write down answers to the following questions at regular intervals. What am I doing? Why am I doing it? How is it going to help me to solve the problem?
5. Leave the problem until the next day. It is surprising how much difference a fresh mind can make.
6. Consult a friend or tutor. Even if your friend is not studying mathematics or one of its applications, simply explaining your problem to someone may help you to gain the insight that you need.

> We can begin to apply these options in turn to our problem. There is first of all no need to check whether we calculated each of the sums to infinity correctly because we abandoned the strategy of guessing altogether. So why not try our second plan of attack? Now the phrase '$35/(1-r)$ not less than 40' can be expressed as the inequality
>
> $$\frac{35}{(1-r)} \geq 40$$
>
> So we can try to solve this inequality as follows:
>
> $\quad\quad \frac{35}{(1-r)} \geq 40$
> $\Leftrightarrow \quad 35 \geq 40(1-r)$; since $1-r$ is positive, as $r < 1$
> $\Leftrightarrow \quad 40r \geq 5$
> $\Leftrightarrow \quad r \geq \frac{1}{8}$
>
> Now this seems far more hopeful as we have a whole range of values of r which will all ensure that the sum to infinity is not less than 40. All we need to do is choose the least value of r from this range, namely $r = 1/8$, and the problem is finally solved. Checking to see that the actual value of the sum to infinity in this case is 40 confirms that our solution is entirely suitable, as the value 40 is certainly not less than 40.

One of the strategies in our list above, or a combination of them, will often lead to the solution of the problem, or alternatively the time allotted to spend on the problem will elapse. In both cases, however, it can be very useful to provide a written summary as to how you tackled the problem. Summarise what you did, why you did it and how it related to the solution of the problem. The most effective way of developing control of your problem-solving is not by an expert telling you what to do, but for you to become aware of how you solve problems yourself.

To illustrate this we can summarise how we tackled our illustration problem. We began by making sure that we *understood* all of the ideas involved in the statement of the problem, so that we would have more chance of solving the problem. The introduction of symbols was particularly helpful as it allowed us to get a handle on

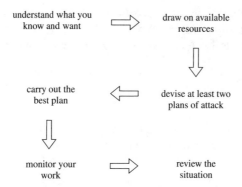
FIGURE 8.1 A STRUCTURED APPROACH TO PROBLEM-SOLVING

the problem. After drawing on a key result about the sum to infinity of a geometric progression, we then *planned* two lines of attack on the problem. The first one we tried failed to work – or at least *monitoring* indicated that it was unlikely to work – but it still taught us the lesson that guessing a solution is not always very effective. The second option, however, led to a solution. Here we can see the importance of gaining understanding, planning and monitoring in the process of solving the problem.

It is worth noting that experts differ from novices in how well they monitor their own problem-solving, in the high proportion of time they spend on understanding the problem and on planning a route to its solution. Following the structured approach outlined in this chapter, as summarised in Figure 8.1 and as employed in the case-studies that follow, will ensure that your problem-solving involves more than just trying the first idea that comes into your head. It will help you to develop greater expertise in problem-solving.

Exercise 8a

1. One plan of attack to solve the problem immediately below would be to measure the required angle. Why is this plan inappropriate and what other ways might there be to tackle this problem?

    Let A, B, C be points on the circumference of a circle, with centre O. Further, let the angle ACB be the angle subtended by the chord AB to the point C and let the angle subtended at the centre of the circle by the arc AB be equal to 40 degrees. Find the angle ACB.

2. Consider the following problem. What resources are relevant to bridging the gap between what is known and what is wanted? Further, outline at least two ways of trying to solve the problem.

   Find the constant force that needs to be applied to a mass of 5 kilograms to enable it to travel a distance of 20 metres in 10 seconds.

3. Carry out each of the plans that you devised in question 2 above. Describe how you monitored the effectiveness of your work.

▶ Case-studies

The case-studies that we will consider in this next section are designed to provide further insight into the structured approach to problem-solving that we have adopted in this chapter.

Acceleration and distance

A vehicle accelerates uniformly from rest at $1.2 \, \text{m/s}^2$ for 12 seconds in one direction. Find how far the vehicle has travelled.

The first stage of our approach is to *understand the problem*. So what does the statement of the problem tell us? We know that a vehicle is accelerating from rest at $1.2 \, \text{m/s}^2$ in a line and that it travels for a period of 12 seconds. Meanwhile, the goal of the problem is to find how far the vehicle has travelled. Attempts to understand the problem will benefit from a clear explanation of each of the ideas involved in what we know and the statement of our goal.

However, it will also be important to employ the skills covered in Part I of the book in order to help understand the problem: for instance, we can introduce symbols that will enable us conveniently to refer to the ideas involved in the statement of the problem. We let a refer to the acceleration, u refer to the initial speed (which in this case is zero), t refer to the time that has elapsed since the vehicle starts travelling, and d to the distance that the vehicle has travelled during time t. So in this case, $a = 1.2 \, \text{m/s}^2$, $u = 0 \, \text{m/s}$, and $t = 12 \, \text{s}$.

The next stage is to *engage in some planning*. How can we use our knowledge that the vehicle uniformly accelerates from rest at $1.2 \, \text{m/s}^2$ for 12 seconds in one direction to find how far the vehicle has travelled? We need to draw on some resources that will allow us to link acceleration in a line, initial speed and time to the distance travelled.

A look through a textbook on mechanics will indicate that there is indeed a formula that links together a, d, t and u (as we have defined them here), namely that

$$d = ut + \frac{1}{2}at^2$$

Alternatively, since acceleration in a line is the rate of change in speed, we may be able to determine the acceleration by differentiating an expression for the speed. The plan we will adopt here is to make use of the formula (although ideally we would also spend some time making sure that we can make sense of the formula itself) given that it seems more straightforward.

So to *carry out our plan*, we can substitute the particular values of u, a and t into our formula to determine the corresponding value of d. Hence we can see that

$$d = 0 \times 12 + \frac{1}{2} 1.2 \times 12^2 = 86.4$$

Thus the vehicle will have travelled 86.4 metres. Clearly, on this occasion the plan has yielded the solution very quickly.

By way of *review* we can say that one key stage of the solution was to make sure that we understood the problem. This initial spadework allowed us to identify a formula that linked acceleration in a line, initial speed and time to distance travelled, and this in turn led to a quick solution of the problem.

> Make sure you spend sufficient time understanding the problem.

Choosing a committee

A committee of three people is to be chosen to represent 11 pairs of twins. If no one pair of twins is allowed on the committee together, how many ways are there of choosing the committee?

One way of solving this problem would be to give each of the members of the group a name and then to simply list all of the possible choices of the committee. This of course is a somewhat unrealistic plan given the number of possible choices involved. A more mathematically satisfying response to the problem involves first of all seeking to gain an *understanding* of the nature of the problem.

The first thing to do, then, is to explore some of the content that forms part of the area of mathematics that is involved: combinatorics. In particular, we need to identify results that relate to this problem. And, as our problem involves different combinations of people on the committee, we will want the following result that determines the number of combinations of r objects chosen from n different objects:

$$^nC_r = \frac{n!}{(n-r)!\,r!}$$

A *plan of attack* on the problem might simply be to start using this above result to help us solve the problem. We could for instance *calculate* the number of ways of choosing three committee members from the 22 people available. But a moment's *reflection* will indicate that many of these combinations will include a pair of twins. How can we rule out these combinations that include pairs of twins? We need a new *line of attack* on the problem.

Looking through established results for inspiration, we can see that if two combinations of different objects cannot occur at the same time (that is, if they are mutually exclusive) then the total number of either one of these combinations is given by adding together the numbers of the two combinations. Can we split up the allowed combinations in our problem into combinations that cannot occur at the same time? This gives us another *way to tackle* the problem, although the solution might still be far from obvious.

One of the most basic features of this problem is that people are split up into pairs. So how can we use this structure to the problem to split the problem into more manageable parts? How can we use what we know in order to carry out our new *line of attack*? Well clearly, in any pair of twins, we can say that there will be an older twin and a younger twin. This *understanding* might help us to rule out a pair of twins from sitting on the committee together.

Of course, experience of solving similar types of problem will also be relevant, and this might also help us to realise that we can consider the number of combinations of any three older twins, two older twins and one younger twin (who is chosen from the remaining nine sets of twins), one older twin and two younger twins (who must be chosen from the remaining 10 sets of twins) and finally any three younger twins. Adding together the numbers of these combinations (because none of them can occur at the same time) will then yield the solution.

We are not finished, however, because we still need to determine the number of each of these combinations. In particular, we need to know how to combine combinations which are independent of each other (as in the case of choosing two older twins, and then choosing one younger twin from the remaining nine sets of twins). When searching through results from this area of mathematics you should be able to find that when combining combinations which are independent of each other, the total number of combinations is given by multiplying together the numbers from the two independent combinations.

All of this leads to the following expression which gives solution of the problem, and which can easily be evaluated using the result given above for ^nC_r:

$$^{11}C_3 + {}^{11}C_2 \times {}^9C_1 + {}^{11}C_1 \times {}^{10}C_2 + {}^{11}C_3$$

In *looking back* at the solution of this problem, we can see the way in which related results helped to shape the way we tackled the solution.

> There is no substitute for familiarity with results from any related areas of mathematics.

Competing populations

The size of a population F is given by F(t) at time t and the size of a population G is given by G(t) at time t, where F(t) and G(t) are determined by the following equations (with a and b constants):

$$\frac{dF}{dt} = at, \quad \frac{dG}{dt} = b$$

Given further that $F(0) = A$ and $G(0) = B$, with constants A and B such that $A < B$, determine the time at which the size of population F equals the size of population G.

How can we try to *understand this problem*? If we restate the goal of the problem, we can say that we need to determine the value of t for which $F(t) = G(t)$. Introducing these symbols allows us to define clearly the problem we are trying to solve. What do we know that will help us to reach what we want? We have two first-order differential equations and an initial condition on each equation.

One *plan* to bridge the gap between what we know and what we want would be to start by solving each of the differential equations, and in order to do this we will of course need to draw on our knowledge of how to solve first-order differential equations. This will give us a rule for each of the functions F and G. We could then use the initial conditions to determine the constant in each of the solutions and plot the graphs of both solutions, before finally reading off from the graph the value of t at which the value of F equals the value of G.

Unfortunately, if we *carry out this plan* – or reflect on it in advance – we will soon find out that the best we can do is attain an approximation to the required value of t. There is no way that we can read off the exact value of t from the graph. An alternative plan is required. Perhaps only once you have solved each of the differential equations, and reached the results that

$$F(t) = \frac{at^2}{2} + A, \quad G(t) = bt + B$$

will it occur to you that by setting

$$\frac{at^2}{2} + A = bt + B$$

we can solve for the required value of t.

Reviewing this outline of the solution, we can say that the insight needed to solve a problem often only occurs when you are part way through a solution.

> Confidence to start work on a problem even if you cannot see the solution is therefore essential.

Extension material: an injective function?

Determine whether or not the real-valued function, g, given by

$$g(x) = (2+x)\sin x$$

is an injective function.

In seeking to *understand* this problem, we need to know what an injective function actually is. Only then will we be able to tell whether the function given in the statement of this problem is actually an injective function. So we can take the following definition of an injective function:

A real-valued function, f, is injective if, for all $a, b \in \mathbf{R}$,

$$f(a) = f(b) \Rightarrow a = b$$

We need to show that our function, g, either satisfies this condition or fails to satisfy this condition. More specifically, the *goal* of our problem is now to decide whether the following statement is true or false:

For all $a, b \in \mathbf{R}$, $g(a) = g(b) \Rightarrow a = b$

One *plan of attack* on the problem would be to try and prove that the statement is actually true. So if we start with the equation $g(a) = g(b)$ can we then show that the equation $a = b$ must also hold? We might try to proceed as follows:

$$g(a) = g(b)$$
$$(2+a)\sin a = (2+b)\sin b$$
$$2\sin a - 2\sin b = b\sin b - a\sin a$$

While it might in theory be possible to apply various double angle formulae or other trigonometric identities, *monitoring* should still indicate that we do not seem to be anywhere near showing that $a = b$.

Given our lack of immediate success, we could alternatively try to show that g is not an injective function. To consider how we might do this, we need to return to the condition itself. For g to be injective, we need the implication to hold for all values of a and b. Hopefully a *plan* will occur to you as a result of this insight: if we find just one value for a and one for b for which it is not true that $a = b$ when $g(a) = g(b)$ then we will have shown that g is not injective.

Perhaps, though, you have reached a grinding halt at this point. There is nothing for it but to try to *understand the problem* more effectively. We can try to apply some of the skills that we explored in Part I of the book. We might look at other examples of injective functions, or consider some particular instances of this function, g, making sure that we understand the ideas involved such as the idea of implication or what a function actually is. Exploring any of these options might well provide the insight needed to complete the problem. If, however, you draw the graph of g (it will be useful for you to employ a graphical calculator or computer algebra system to do this) and look at some of the details of this graph, you will be able to see when $x = 0$, $x = \pi$ and $x = 2\pi$ that the value of g equals zero. But this is all the insight that we need to solve our problem, because we have found several different values of a and b which give the same values of $g(a)$ and $g(b)$: for instance, $g(0) = g(\pi)$, but $0 \neq \pi$. Hence we have shown that g is not injective.

In conclusion we can *review* our work to see that these different phases of understanding the problem, devising a plan, carrying out the plan and looking back all contributed to the final solution of the problem.

> Make sure that your problem-solving is similar to the problem-solving of an expert.

Exercise 8b
Using the approach indicated in this chapter, solve each of the following problems. Further, add a commentary as to how you went about solving each problem.

1. Someone drives a vehicle at one speed for three hours and then at a different speed for another hour, covering a total distance of 66 miles. On another journey the vehicle is driven at the first speed for two hours and at the second speed for 5 hours. This time the vehicle covers 200 miles. Determine the two speeds.

2. The hypotenuse of a right-angled triangle is twice the length of one side. If the area of the triangle is $\sqrt{2}$ cm² then find the lengths of each of the three sides of the triangle.
3. Find a problem of your own choosing, perhaps one that you failed to solve on a previous occasion. Apply the approach outlined in this chapter to your selected problem.

Extension Exercise 8

1. Choose a given area of your study in mathematics or its applications. Solve a number of problems in the area. Can you characterise the types of strategy that you used to solve these problems?
2. Consider a group with elements x and y both of order 3, and satisfying $x^2y = yx$. Show that $yx^{-1} = xy$. Further, provide a commentary outlining how you tried to solve this problem and explaining why you proceeded as you did.

Reflection

- How does your own approach to solving problems compare with the approach outlined in this chapter?
- How much attention do you pay to understanding, planning and monitoring in your problem-solving?

Summary

When solving a problem: understand the problem; engage in planning; carry out a plan; review the situation.

9 Applying Mathematics

> This chapter aims to:
>
> enable you to apply mathematics to the real world.

▶ Why apply mathematics?

Imagine two ancient herdsmen bargaining with each other over an exchange of cattle. Who got the better deal would be a matter of poverty and riches. By representing concepts from the real world, such as 'the size of a herd', with mathematical ideas, such as 'number', our early ancestors were able to use mathematics to manage their environment.

The primary motivation for applying mathematics today remains that of utility. Mathematical representations of the real world allow us to explain how the world operates, enabling control, optimisation of performance, prediction and design. Newton's law of gravitation, for instance, uses mathematical ideas to model the way in which different objects or bodies are attracted to each other, and that knowledge is of course very useful when you are trying to launch a space shuttle.

It is hard for us to realise now how much our world has actually been shaped by the application of mathematics. Just think what life would be like without electronics, modern transport, telecommunications and information and communication technology. All of these have resulted from the application of mathematical ideas. Indeed, the success of mathematics in underpinning scientific explanations of the world has led to applications of mathematics in a huge range of areas of life. Politicians use performance indicators to measure how effectively government departments are providing public services. Athletes maximise their levels of fitness by monitoring their bodies. Some mothers even aim to feed their babies at regularly spaced intervals.

Finally, it is worth observing that the ability to apply mathematics is a highly-valued skill. Given the great benefits that mathematics has brought to our world, it is not surprising that those students who have mastered mathematics and its applications will find their skills in demand. The aim of the rest of this chapter is to help you make sure that you are one of these students.

> Application of mathematics helps us to explain and control the real world.

▶ How to apply mathematics?

The fundamental idea involved in applying mathematics is the idea of a model. We can say that a street map of a town is a 'model' of the town. The model provides an accurate representation of the lengths of the streets and of the connections between them (see Figure 9.1). By contrast, other features of the town, such as the heights or colours of the buildings, are not represented on the map.

As far as applying mathematics is concerned, the models we deal with are mathematical models of different aspects of the real world.

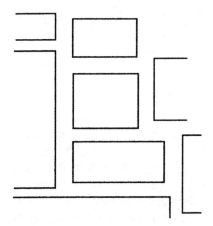

FIGURE 9.1 A TYPICAL PORTION OF A STREET MAP

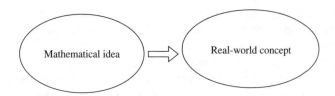

FIGURE 9.2 MATHEMATICAL IDEAS REPRESENT REAL-WORLD CONCEPTS

In particular, individual mathematical ideas are taken to represent given concepts from the real world, as illustrated in Figure 9.2. Typically, the real-world concepts involved are each represented by a mathematical variable, and this usually means that we can measure each of the concepts: for example, we use the idea of a number to represent the concept of distance or the idea of a vector to represent the concept of acceleration.

A mathematical model might then go on to represent relationships between concepts from the real world in terms of relationships between mathematical ideas. For instance, Newton's second law of motion is given by the well known equation:

$$\mathbf{F} = m\mathbf{a}$$

Here the concept of the force acting on a body is represented by the vector $\mathbf{F}$, the concept of the acceleration of the body is represented by the vector $\mathbf{a}$ and the concept of the body's mass is represented by the real number m. The equation itself then provides a mathematical representation of the way in which these concepts of force, acceleration and mass are related to each other.

> A mathematical model is a mathematical representation of a conception of some aspect of the real world.

Once a model is in place we can then use it to provide insight into the particular aspect of the real world that is being modelled, and to solve related problems. It might in fact be very useful to know that, in order to accelerate a motorbike 'from 0 to 100 mph in 5 seconds', you need to exert a certain force on the motorbike.

It is, however, the case that some models are better than other models. Mathematical models are not either right or wrong: they are simply either less or more effective at explaining events in the real world. It is, of course, possible to model the surface of the earth using the geometrical idea of a plane, and 'Flat Earth' societies exist to promote this particular model. For an architect, however, this model of the surface of the earth might be perfectly adequate.

Clearly, in order to apply a model you need a good understanding of both the model itself and of the aspect of the real world that is being modelled, and there is no better way of reaching this understanding than appreciating the way in which the model was created in the first place. This allows you to appreciate a model's strengths and weaknesses. It also allows you to adapt existing mathematical models and to create your own.

Creating models of the real world

The process of creating a model of some aspect of the real world is best described as a cycle, called the *modelling cycle*. A version of this cycle is outlined in Figure 9.3.

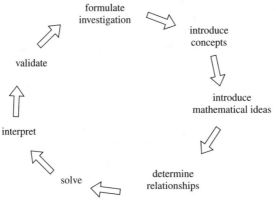

FIGURE 9.3 THE MODELLING CYCLE

> In what follows we will illustrate each of the stages with reference to creating a model of a simple problem from economics.

1. Formulate the investigation The first stage of the modelling cycle is to formulate a *clearly defined* investigation. For instance, there is is little point trying to model something to do with a company making a profit. We need to be clear exactly which aspect of the company's business we wish to investigate.

> To illustrate the modelling cycle we will investigate how much you should charge your customers for a particular piece of merchandise if you wish to maximise your profit. One way of ensuring that you are clear about the nature of the problem is to restate the investigation in your own words. For this problem from economics we might thus ask if there is any way we can tell how the price of the merchandise will affect the actual demand for it.

2. Introduce concepts The next stage of the cycle is to *identify concepts* that help us to make sense of the investigation. At this stage of the modelling cycle the focus is on the most relevant concepts, and identifying which concepts are the most relevant clearly involves making assumptions about what is important in the problem. Of course, some of these assumptions may prove inappropriate, but later stages of the cycle will check this for us.

In many fields the important concepts will already have been identified. One of the main reasons why Newton was able to model so effectively was that earlier thinkers had already identified many of the key concepts. Indeed, Newton himself said that he had only been able to see as far as he had done because he had stood on the shoulders of giants.

> For our investigation, the main concepts have already been defined by economists. There is first of all the idea of a 'good that is demanded': something that people are willing to go to some lengths to acquire; we can then qualify this by the concept of 'how much of a good is demanded'; and we also have the idea of the 'price' at which someone offers a good for sale.

In all of this, it is critical to ensure that you understand the concepts involved. One way of checking whether you do actually understand the concepts is to provide a range of examples for each concept, as we saw in Chapter 2.

> So in our investigation we might be talking about a good such as a photocopier or a pound of carrots.

3. Introduce mathematical ideas Each of the concepts that we have introduced in the previous stage of the cycle should now be *represented* by a specific mathematical idea. We could represent the quantity of a good demanded by any of the numbers from the set $\{0, 1, 2, 3, 4, \ldots\}$ but in order to make the mathematics more convenient we use the idea of any positive real number instead.

> In this case the concept of the quantity of a good demanded is represented by the idea of a real variable. Of course some goods, such as photocopiers, may only come in whole numbers while other goods, such as carrots, can be bought in fractions of a pound. But in modelling we usually need to sacrifice some precision for the sake of mathematical convenience. Similarly, we can also represent the price of the good by making use of real numbers.

Furthermore, you need to make it clear which concepts have been identified with which mathematical ideas and, as we saw in Chapter 7, the easiest way to do this is to give the concept and the mathematical idea the same label.

> So on this occasion we label the quantity demanded of a good D and the price charged for the good P.

4. Determine relationships The next stage is to *determine any relationships* between the mathematical ideas. Usually you will need to concentrate on the most important relationships as we are interested in the major factors in the behaviour of a system. You will often find that others before you have already formulated laws involving the concepts, and hence you need a good understanding of existing theory before starting to model. This approach to modelling is called *theory-driven modelling*.

> For our example, there is a basic economic law which states that demand falls as price increases. This assumes that the demand for a good is dependent upon its price, so we can represent the relationship between demand and price using the mathematical idea of a function. We say that
>
> $$D = f(P)$$

> **Box** *Continued*
>
> where f is a function that associates values of the real variable D with values of P. Furthermore, we can assume for the sake of simplicity that the function is linear. So we obtain the following equation as a model of the relationship between the quantity demanded of the good D and the price of the good P:
>
> $$D = aP + b, \quad a < 0$$
>
> Here, the constant a determines how quickly demand falls as price increases and the constant b determines the level of demand for a good that is offered without cost to consumers. These two constants can be adjusted depending on the particular good being considered.

An alternative to starting with theory would be to collect a set of data detailing the quantity of a good demanded during different periods of time and to see if there is any connection with the actual price. This approach to determining relationships between the concepts is called *data-driven*, and is usually more challenging than theory-driven modelling.

You further need to make sure that you understand the relationships involved.

> What, for instance, does it mean for demand to fall as price increases? And what is a function? It is worth considering examples and particular instances. For instance, we could consider some particular instances of the relationship between price and demand. What is the demand equal to if the price is set at 0?
>
>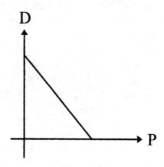
>
> **FIGURE 9.4** A GRAPH OF DEMAND AGAINST PRICE

In addition, a diagram such as Figure 9.4 may be useful. In spelling out the connection at the global level between this graph and our model we can see that the line has a negative gradient: that is, the graph depicts the assumption that if the price increases then the demand for the good will decrease.

5. Use the model to solve the problem Many students tend to focus on the fifth stage of the modelling cycle. This is the stage in which you *explore* the implications of the mathematics and *solve* mathematical problems. All the power of mathematical thought is now at your disposal. Indeed, the mathematical exploration may lead us to discover new relationships between the concepts and even lead to the introduction of new concepts.

The mathematics involved in the case we are considering of price and demand is, however, straightforward. So in this case we are not so much solving a set of equations as using our mathematical formulation to provide further information about how the variables are related to each other. For instance, in our problem about supply and price we can explore the mathematics as follows:

$$D = aP + b \Leftrightarrow D - b = aP \Leftrightarrow P = \frac{1}{a}D - \frac{b}{a}$$

Here, the mathematics indicates that we can also define P as a function of D.

6. Interpret Once the mathematical representation of the problem has been fully explored, it is important to *interpret* the results in terms of the real-world situation. This is to answer the questions 'What does the solution say about the way in which the concepts are related to each other?' and 'How is the real world connected to the mathematical world you have just created?'

For instance, in our example we saw that the mathematics indicated that price is also dependent on demand. So if more customers are trying to purchase a particular good than a supplier can cope with, then the supplier can increase the price until the

> **Box** *Continued*
>
> demand has fallen to a level which that supplier can easily meet. We can also interpret some individual particular instances. For instance, the graph might indicate that if the price is set at £250 then customers in a given market will demand 25 013.47 photocopiers. Clearly, however, it will not be possible to produce 0.47 of a photocopier, so any interpretation that we make needs to take account of the assumptions made in setting up the model. In this case the simplest option is to approximate the number of photocopiers with a whole number: 25 013.

7. Validate The final stage is to *validate* whether the model actually matches the real-world situation. In validating you are checking to see whether your work has any genuine connection with reality, so it is worth making predictions from the model and comparing them with the actual observations of the real-world problem. It is also worth checking to see whether outputs from the model change as inputs change in ways that you would expect.

Validation, however, may indicate that the model needs to be revised. In this case you will need to return to the first stage of the modelling cycle. Should the problem be re-formulated in terms that are more amenable to a mathematical investigation? Have we neglected any important concepts?

> It is in fact clear that the way in which our model predicts that price is related to demand does not actually match very well what happens in practice. The relationship between price and demand is in reality more complicated than the model allows. For instance, a variety of other factors will also affect the demand for a good rather than just its price. These factors will include the amount spent on advertising the good, the income of the consumers, the price of any substitutable goods which might be bought instead of the good, and so on. A more accurate model would need to take account of such factors.

> Mathematical modelling encompasses far more than just the solution of a mathematical problem.

Using models of the real world

We saw in the introduction to this chapter that we apply mathematics in order to understand and control the real world. And we have seen in the last section that we primarily apply mathematics by creating mathematical models. It is now worth thinking about how we might use a model.

In using a model, we first of all need to be aware of its limitations. In particular, we need an awareness of the assumptions that originally went into making it.

> Any mathematical model only applies in a limited set of circumstances. Models must not be employed for uses for which they are not designed.

> For instance, if we are working in a commercial environment in which it is known that there are several alternative goods which consumers could equally well buy, then we would need to recognise that our earlier model of the relationship between price and demand is unlikely to be of much use. We would need to employ a more sophisticated model.

However, we also need to be aware of the limitations of the modelling process as a whole. All mathematical models require that concepts from the real world are formulated in ways that are amenable to mathematical representation. This means that the focus is on what can be counted or quantified, but there may be important concepts that cannot be quantified at all readily.

> A mathematical model is not reality, it is only a model of a particular conception of reality. The real world always ultimately escapes our mathematical models of it.

For instance, medical need may be assessed on the basis of a numerical scale, but how is the scale determined? Is there any scope for concepts such as dignity and compassion? The very reliance on quantification and measurement affects our judgement.

> The uses to which you can put a model depend on the limitations of both the individual model concerned and the whole modelling process.

Once we are aware of the limitations of modelling, there are of course a whole variety of uses to which we can put models. As noted in the introduction to this chapter, we can use models to explain how the world operates, enabling control, optimisation of performance, prediction and design. A good model can reduce the need to carry out experiments, it can aid the development of theory or it can test whether a theory is correct. The particular use to which you put any model will of course depend on what you want to achieve and on the real-world situation concerned. What you should not do is just concentrate on the mathematical model itself.

Exercise 9a

1. What assumptions are involved in representing the real-world concept of an incline by the mathematical idea of a line with a specified gradient?
2. Why might Euclid have proposed that light is always propagated in straight lines?
3. Provide an interpretation of the following mathematical relationship in terms of relationships between the relevant real-world concepts:

    $\rho = m/V$ where ρ is the density of an object, m is its mass and V is its volume.

4. Explain the limitations of using the number of steps that someone would take to walk to an object to model the distance between the person and the object.

▶ Case-study: a model of a lever

In this case-study we consider a mathematical model of a lever. Now a lever is perhaps the world's oldest machine. It enables a

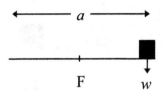

FIGURE 9.5 A MODEL OF A LEVER

force applied at one point effectively to act at another point. But how does the lever work, and how can we know exactly how much force we need to apply to the lever in any given situation?

To formulate our investigation more precisely, we can say that we aim to understand the behaviour of the lever when two or more weights are placed at different points along its length. In particular, we are interested in predicting when the lever will remain in equilibrium.

Now we will consider a lever as an object, such as a bar of a certain length whose centre rests on a fulcrum. We can represent the bar by a line, of length a (where a is a positive real number) in a given measure and the fulcrum by a point, labelled F, at the centre of the line. For convenience we assume that the bar is weightless. A weight is then represented by a vector of a given magnitude, w, acting downwards at the point along the line where the weight has been placed. This provides the basic mathematical model of the lever (bringing us up to stage 3 of the modelling cycle), as pictured in Figure 9.5.

In trying, however, to understand the behaviour of a lever, we can further draw on two fundamental assumptions. First of all, we can assume on the basis of experience that if three weights all of magnitude w are placed one at either end of the bar and one at the centre then the lever will not move, but will remain in *equilibrium* (Figure 9.6a). We can also assume from symmetry that if the weights at one end of the bar and at the centre are replaced by a single weight, which is twice as heavy, at a point half-way between the centre and the end of the bar then the lever remains in equilibrium (Figure 9.6b). We can conveniently represent this latter instance of a system in equilibrium by the equation

$$w \times \frac{a}{2} = 2w \times \frac{a}{4}$$

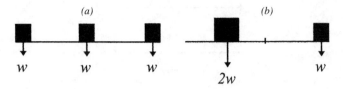

FIGURE 9.6 MODELS OF A LEVER INDICATING TWO FUNDAMENTAL ASSUMPTIONS

So not only do we have a mathematical representation of a lever, we also have some knowledge about the way in which the different elements of the system are related to each other (stage 4 of the modelling cycle). Now is the time to explore the mathematical consequences by working through stage 5.

Take a lever (now of length $4l/3$) with weights of magnitude $3w$ and $2w$ placed as indicated in Figure 9.7a. Then by our argument from symmetry we can replace the weight of magnitude $3w$ with three separate weights, each of magnitude w and placed as indicated in Figure 9.7b. And we can again replace two of these three weights with a single weight as indicated in Figure 9.7c. Now according to a version of the first of the above two fundamental assumptions, the system in Figure 9.7c is in equilibrium. But this system is equivalent to the system in Figure 9.7a, which is thus also in equilibrium. We can represent this instance of equilibrium by the equation

$$3w \times \frac{l}{3} = 2w \times \frac{l}{2}$$

Now it is possible to employ a similar argument to show that the system represented by the equation

$$\alpha w \times \frac{1}{\alpha} = \beta w \times \frac{1}{\beta}$$

is also in equilibrium. In this system a weight of magnitude αw is placed at a distance $1/\alpha$ from the fulcrum with the weight βw at a distance $1/\beta$.

Furthermore, we can see that the magnitude of the weight multiplied by the distance from where the force acts to the fulcrum is an important variable in determining whether the system is in equi-

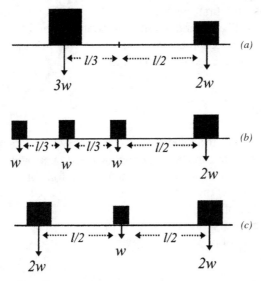

FIGURE 9.7 MODELS OF A LEVER IN EQUILIBRIUM

librium. This allows us to define the idea of the *moment of a force* about a point as the magnitude of the force multiplied by the distance from the point.

In interpreting the model as far as real levers are concerned we can now appreciate that a heavy weight on one side of the lever will be balanced by a lighter weight on the other side if the lighter weight is further away from the fulcrum. It is also possible to validate this model of a lever. We can easily predict from our mathematical model whether or not a lever will be in equilibrium depending on the magnitude and location of any weights.

Finally, it is worth making a few comments on how to put our model to use. It might be very useful to be able to predict how much force to apply at one end of a lever in order to balance a load at the other end, if only to be able to play a children's game such as 'seesaw'. But the model does have its limitations. Clearly if you were to use for the lever a bar that weighed more on one side than the other, then our assumption that the bar was weightless would cause difficulties.

In conclusion we can note that we have seen here an example of the power of mathematical thought. We have both reached a new law that describes the behaviour of the lever and have introduced a new concept (the idea of a moment of a force about a point). Archimedes was the first person to explore the mathematical consequences of this mathematical representation of a lever. His work provides a truly acclaimed example of the way in which the natural world can be modelled by mathematics.

Exercise 9b

1. Drawing on the modelling cycle and on the skills covered in Part I, as demonstrated in the previous case study, provide an explanation of a standard model of each of the following concepts. (If necessary draw on a relevant textbook.)
   (a) Velocity of a body.
   (b) The central tendency of a set of data.
   (c) A model of population growth.

Extension Exercise 9

1. Drawing on the modelling cycle and on the skills covered in Part I, as demonstrated in the previous case study, provide an explanation of each of the following models. (If necessary draw on a relevant textbook.)
   (a) A model of both demand and supply.
   (b) Newton's model of planetary motion.
   (c) The Poisson model.
2. Provide a full explanation of each of the following statements.
   (a) The rate of photosynthesis in relation to light intensity is modelled by a mathematical function.
   (b) The marginal revenue is the derivative of the total revenue with respect to demand.
3. Identify the real-world concepts involved in the following equations and specify how they relate to each other.
   (a) The Boltzmann equation.
   (b) The equation that specifies Hooke's Law.

> **Reflection**
> Upon which stage or stages of the modelling cycle do you tend to focus in your study?

Summary

- To apply mathematics you create a mathematical model of some aspect of the real world.
- A mathematical model involves formulating an investigation, introducing both concepts from the real world and mathematical ideas, determining relationships, finding a mathematical solution, interpretation and validation.

10 Constructing Proofs

> **This chapter aims to:**
>
> enable you to understand how to establish the truth of statements.

▶ The need for understanding

One of the most fundamental of all mathematical tasks is to establish that a given statement is in fact a true statement. We do this by means of a *proof*, a correct logical argument which demonstrates the truth of the statement as long as the initial assumptions were true. Indeed, we saw in Chapter 6 how important it is that truth in mathematics is established on a secure basis. But how often have you tried to construct a proof of some result and found yourself hopelessly lost?

One way to proceed is to memorise the proofs of any standard results you might be expected to reproduce in an examination. But it is difficult to memorise more than a few proofs for any one occasion and sometimes you might be asked to prove a result which you have not seen before. Moreover, it would not be easy for you to use a result for which you have simply memorised the proof.

The alternative, which we will explore in this chapter, is to make sure that you understand any proofs you are expected to know, so that you can reconstruct and use them at will. Then, if you are asked to prove a result you have not already seen, your understanding of what makes for a valid argument will be of great value.

There are three main strategies that we will consider to develop an understanding of a proof. These are outlined below. Notice how the last two of these strategies draw on skills of synthesis and analysis.

> **In order to understand a proof**
>
> 1. Understand what you are trying to prove.
> 2. Gain an overview of the proof.
> 3. Make sure that you understand the details of the proof.

We will apply this threefold approach to proofs of what mathematicians call *theorems*. A theorem is a statement which can be proved true. For instance, the statement

$x^2 + y^2 = z^2$, where x, y and z are real variables which specify the three lengths of the sides of a right-angled triangle

is a theorem (called 'Pythagoras' Theorem') because a proof exists to show that it is true.

It would also be possible to regard the statement '$3^2 + 4^2 = 5^2$' as a theorem because it can be proved that this statement is true. But in practice, mathematicians tend to use the word 'theorem' to refer to more general statements that hold true in a variety of cases. Instead, we would regard the statement '$3^2 + 4^2 = 5^2$' as a particular instance of Pythagoras' Theorem. It is also possible to use the term 'proof' in connection with proving that you have found the correct solution to a problem, but we will focus in this chapter on making sense of proofs of theorems.

▶ Understand the theorem

The primary way in which we will seek to understand what we are trying to prove is to apply the skills of Part I of the book. After all, these are some of the fundamental ways in which mathematicians try to gain understanding of the ideas with which they work, but it will also be important to gain an appreciation of what any given theorem is useful for. This will be of particular interest to readers whose study focuses on applications of mathematics.

Employ the skills

We have already noted above that a theorem is usually a general statement. This provides us with a natural opportunity to consider the first of our skills from Part I: the use of *examples*. In this case we will consider different examples of the theorem. The challenge, then, is to create a varied set of examples.

> To illustrate some of the points made in the rest of this chapter, we will consider the following well-known theorem:
>
> *Theorem*: Given a real variable x and the real numbers a, b, c for which $a \neq 0$ and $b^2 - 4ac \geq 0$, then:
>
> $$ax^2 + bx + c = 0 \Leftrightarrow x = \frac{-b \pm \sqrt{b^2 - 4ac}}{2a}$$
>
> We can then see in Figure 10.1 some different examples of the theorem. It should be fairly obvious why the simple example is a simple example, and why the typical example is typical, but it might be less obvious to some readers why it is claimed that the unusual example is unusual. It is because in this example there is only one value of x that satisfies the equation, rather than the more usual two values which are evident in the other examples in the table. You should also be able to see from Figure 10.1 that the theorem itself is a *generalisation* of these examples. Now this does not guarantee that the generalisation is valid, and thus that the theorem is valid, but it should help you to understand what the theorem actually claims.
>
> | Initial example | $2x^2 + 3x + 0.5 = 0 \Leftrightarrow x = \dfrac{-3 \pm \sqrt{9 - 4 \times 2 \times 0.5}}{2 \times 2}$ |
> |---|---|
> | Simple example | $x^2 - 1 = 0 \Leftrightarrow x = \dfrac{0 \pm \sqrt{0 - 4 \times 1 \times (-1)}}{2 \times 1}$ |
> | Typical example | $-3x^2 + 2x + 0.6 = 0 \Leftrightarrow x = \dfrac{-2 \pm \sqrt{2^2 - 4 \times (-3) \times 0.6}}{2 \times (-3)}$ |
> | Unusual example | $x^2 + 2x + 1 = 0 \Leftrightarrow x = \dfrac{-2 \pm \sqrt{4 - 4 \times 1 \times 1}}{2 \times 1}$ |
>
> **FIGURE 10.1** A COLLECTION OF EXAMPLES OF THE THEOREM

If the theorem involves ideas that can be represented spatially then it will be important to include a *diagram* in your quest to understand the theorem. Details or an overview of the diagram can be linked to the statement of the theorem.

> In this case we can plot in Figure 10.2 the graph of the real-valued function given by $f(x) = 1 - x^2$. Looking at some of the details of our graph, we can see that the function assigns to each of the real numbers 1 and -1 the real number 0. If we connect these details with the theorem itself, we can see that when the value $x = 1$ is substituted into the quadratic equation then the equation holds and that $x = 1$ is one of the two values specified on the right-hand side of the equivalence sign in the statement of the theorem. A similar connection is evident with the value $x = -1$.

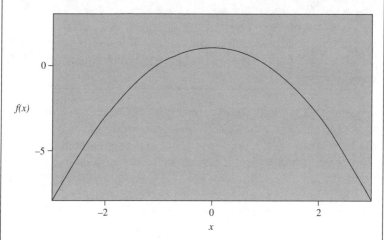

FIGURE 10.2 THE GRAPH OF THE REAL-VALUED FUNCTION GIVEN BY $f(x) = 1 - x^2$

The next skill involves focusing on the *symbols* used in the statement of the theorem. Remember, that means not only must you have seen each symbol before and know how to pronounce it, but you must also be able to provide a full explanation of the meaning of each symbol.

> So let us take each of the symbols from the theorem in turn. We start with the real variable x, an idea which every reader should now fully understand. But what role do the real numbers a, b and

> **Box** *Continued*
>
> c play? These numbers are unspecified real numbers, but we refer to them as *parameters* rather than variables. This is because in any one example of the theorem they each have one fixed value, unlike the variable x which can take on more than one value within a single example of the theorem. Finally, one further symbol – $\Leftrightarrow$ – particularly stands out.
>
> This symbol refers to a logical idea that was covered in Chapter 6: the idea of 'equivalence'. At this point, readers might find it helpful to re-read the relevant section in Chapter 6. Saying that statements A and B are equivalent to each other means that if A is true then B is true, and if B is true then A is true. And it also means that if A is false then B is false and if B is false then A is false. In the context of two algebraic equations being equivalent to each other, this means that the values of x which satisfy one equation are exactly the same values of x which satisfy the other equation.

In general, it is usually worth looking at the *logical* structure of the theorem itself. Now you might think that a discussion of logic would only be relevant to the proof itself, but statements themselves can incorporate further logical ideas, as we have just seen in the above paragraph.

It is also worth noting that spelling out the meaning of the symbols in a theorem often overlaps with an *analysis* of the more basic ideas that are involved. Given that the whole of our theorem is expressed in symbols this should not be so surprising. All you need to do in this case to analyse the theorem is to list the ideas that are represented by the symbols.

Finally, for any example of the theorem it may be possible to consider some *particular instances*. Given the complexity of most theorems, it is often important to focus on different cases of an example of the theorem.

> To illustrate this we can pick out the simple example in Figure 10.1:
>
> $$x^2 - 1 = 0 \Leftrightarrow x = \frac{0 \pm \sqrt{0 - 4 \times 1 \times (-1)}}{2 \times 1}$$

We have waited to discuss particular instances of this example until now since we will make use of the above discussion on the meaning of the equivalence of two statements. We can take as particular instances different values of x. For $x = 2$ we can see that the statement becomes

$$4 - 1 = 0 \Leftrightarrow 2 = \pm 1$$

Clearly both of these equations are false statements, so it is true to say that they are both equivalent to each other. And if $x = 0$ then they are again both false, so the statement as a whole is true. But if we take $x = 1$ then it is true to say that $1 - 1 = 0$ and also true to say that $x = 1$ or $x = -1$. These particular instances help us to see that this example of the theorem is valid whatever value of x is chosen (although on their own they do not prove this to be the case).

> Theorems need to be 'unpacked' if you are to fully understand them. The skills of Part I help you to do this.

Using theorems

Would you claim that you understood a word if you only knew its dictionary definition but could not use it in speech or writing? The same point can be made about understanding a theorem. To really understand a theorem you need to be able to make use of it.

> So how can we make use of our theorem on the solutions of a quadratic equation? Well, the theorem enables us to say that whenever we have a quadratic equation for which certain conditions hold, then we automatically know the values of x which ensure that the equation is true. We do not need to do any more work to find out these values, the theorem simply tells us what they are.
>
> For instance, we can take the equation: $x^2 - 5x + 4 = 0$, where x is a real variable. Now it is the case that 1, -5 and 4 are all real numbers, $1 \neq 0$ and $(-5)^2 - 4 \times 1 \times 4 \geq 0$, so we know from our theorem that the solutions of the equation must be given by:
>
> $$x = \frac{5 \pm \sqrt{25 - 16}}{2}$$

> **Box** *Continued*
>
> Without doing any more work, we know that if we substitute these two values of x into the equation $x^2 - 5x + 4 = 0$ then they will satisfy it. And we also know that they are the only values of x that will satisfy the equation.

> Once a theorem has been proved, then it applies whenever the conditions of the theorem are met.

Much of the power of mathematics in fact comes from theorems. All you need to do is prove once and for all that the theorem is true, and then you can use the theorem as freely as you wish because you know that it must be true.

Exercise 10a

1. Employ the skills from Part I of the book to understand the following theorem:

    For positive real numbers x and y: $\log_e x + \log_e y = \log_e xy$.

2. The Fundamental Theorem on Equivalence Relations states the following:

    If R is an equivalence relation on a set S then the distinct equivalence classes partition S.

    So if we take some particular equivalence relation called P on a set X then what can we say about its equivalence classes?

3. Describe a use of Pythagoras' Theorem.

▶ See the overall structure

Have you ever been able to follow each line of a proof but still felt that you did not understand why the proof works? There is clearly a great deal more to understanding a proof than being able to make sense of each of its lines.

> Take for instance the proof of our theorem on the solutions of a quadratic equation. We can outline the proof of the theorem (alongside a re-statement of the theorem itself), as follows:
>
> *Theorem*: Given a real variable x and the real numbers a, b, and c for which $a \neq 0$ and $b^2 - 4ac \geq 0$, then:
>
> $$ax^2 + bx + c = 0 \Leftrightarrow x = \frac{-b \pm \sqrt{b^2 - 4ac}}{2a}$$
>
> *Proof*:
>
> $$ax^2 + bx + c = 0 \quad (1)$$
> $$\Leftrightarrow \quad x^2 + \frac{bx}{a} + \frac{c}{a} = 0 \quad (2)$$
> $$\Leftrightarrow \quad \left(x + \frac{b}{2a}\right)^2 - \left(\frac{b}{2a}\right)^2 + \frac{c}{a} = 0 \quad (3)$$
> $$\Leftrightarrow \quad \left(x + \frac{b}{2a}\right)^2 = \left(\frac{b}{2a}\right)^2 - \frac{c}{a} \quad (4)$$
> $$\Leftrightarrow \quad x + \frac{b}{2a} = \pm \frac{\sqrt{b^2 - 4ac}}{2a} \quad (5)$$
> $$\Leftrightarrow \quad x = \frac{-b \pm \sqrt{b^2 - 4ac}}{2a} \quad (6)$$
> $$\text{Hence} \quad ax^2 + bx + c = 0 \Leftrightarrow x = \frac{-b \pm \sqrt{b^2 - 4ac}}{2a} \quad (7)$$
>
> It is relatively easy to see, for instance, why line (6) of the proof follows from (5). All we have done is subtract $b/2a$ from both sides of the equation.

But even if we can make similar sense of each line of the proof, this still does not give us a feeling for why the proof as a whole is valid. What we need to do is gain an understanding of the overall structure of the proof. We will look at two ways to help us do this.

It is first of all important to make sure that you can understand the *type of proof* employed. The focus here is on the logical structure of the proof. Why is the proof a correct logical argument?

> We can call the proof of our theorem outlined above a *direct proof* or a proof by means of equivalence. The two equations in the

> **Box** *Continued*
>
> theorem are equivalent to each other because they are linked together by a chain of equivalent statements. Hence if the first equation holds then the next equation holds, and so on until the final equation holds; and if the first equation is false then the next equation in chain is false, and so on until the final equation is also false.

In addition to understanding the logical structure of a proof, it is also important to provide a more intuitive overview or summary of the proof. This can help us feel that we understand where the proof is going.

> In the case of our theorem we can say that the proof in some sense transforms the first equation into the second equation by making moves that are all allowed by the laws of arithmetic. The drive behind these moves is to reach a situation in which the variable x is on one side of the equation and the parameters and other real numbers are on the other side of the equation.

> To understand a proof:
>
> Developing a feel for the logical structure of the proof and providing a clear summary of the proof leads to the kind of insight that is needed for you to feel that you understand a proof.

Exercise 10b

Outline the logical structure of the given proof of each of the following two theorems. Further, provide a short summary of each proof.

1. *Theorem:* $a^2 = b^2 \Leftrightarrow a = b$ or $a = -b$

    Proof:

    $$
    \begin{align}
    & a^2 = b^2 \tag{1}\\
    \Leftrightarrow\ & a^2 - b^2 = 0 \tag{2}\\
    \Leftrightarrow\ & (a-b)(a+b) = 0 \tag{3}\\
    \Leftrightarrow\ & a = b \text{ or } a = -b \tag{4}\\
    \text{Hence}\ & a^2 = b^2 \Leftrightarrow a = b \text{ or } a = -b \tag{5}
    \end{align}
    $$

2. *Theorem*: For positive real numbers x and y, $\log_e x + \log_e y = \log_e xy$.

   Proof:

   $\quad$ Let $\log_e x = a$ and let $\log_e y = b$. $\hfill (1)$

   $\quad$ By definition of $\log_e$, $e^a = x$ and $e^b = y$. $\hfill (2)$

   $\quad$ Therefore $\quad\quad\quad e^a \times e^b = xy \hfill (3)$

   $\quad\Leftrightarrow \quad\quad\quad\quad\quad\quad e^{a+b} = xy \hfill (4)$

   $\quad\Leftrightarrow \quad\quad\quad\quad\quad a+b = \log_e xy \hfill (5)$

   $\quad\Leftrightarrow \quad\quad\quad \log_e x + \log_e y = \log_e xy \hfill (6)$

   $\quad$ Hence $\quad\quad \log_e x + \log_e y = \log_e xy \hfill (7)$

▶ See the details

It is, however, still important to make sure that you can make sense of each line of a proof. You will of course need to employ the skills covered in Part I to aid your understanding of each line. But otherwise, an efficient way of developing a detailed level of understanding is to add a commentary to each line of the proof. The rest of this section will focus on how to add this commentary, given that we have already had plenty of practice in employing the skills of Part I.

You should look to add comments to each line of the proof according to the following three basic principles.

Comments may relate to how the line:

- follows from some earlier lines;
- draws on other mathematical ideas and results;
- fits into your overview of the proof.

We can take each of these types of comment in turn. The most obvious comments to make first of all are those that relate the line to the previous lines of the proof. How does this line follow from the earlier lines of the proof?

For our theorem, for example, we can see that line (3) follows from (2) because the following equations hold:

$$\left(x+\frac{b}{2a}\right)^2 - \left(\frac{b}{2a}\right)^2 = x^2 + \frac{bx}{a} + \frac{b^2}{4a^2} - \frac{b^2}{4a^2}$$
$$= x^2 + \frac{bx}{a}$$

It can also be important to add in comments that relate to how the statement draws upon ideas from elsewhere. Given how mathematical ideas are linked to other mathematical ideas it should not be surprising that any given proof will usually draw on a range of other ideas and results.

In the case of our theorem, we can see that line (2) follows from (1) as follows. We multiply each side of equation (1) by the expression $1/a$, which we are allowed to do because $a \neq 0$. We can then see that

$$\frac{1}{a} \times (ax^2 + bx + c) = \frac{ax^2}{a} + \frac{bx}{a} + \frac{c}{a}$$

as a result of the distributive law of arithmetic.

You should finally be able to relate all of the details to the overall structure. The aim is to indicate exactly how each line is related to the informal overview and to the logical structure of the proof. It is not enough to vaguely claim that the detail is somehow related to the overview; only by connecting the details of the proof to the structure are you likely to grasp the full meaning of the details.

So again, in the case of our theorem we can note that each of the equations in lines (1)–(6) provide one link in the chain of equivalent statements which characterises the logical structure of the proof. And we can also see that to transform one line into the next line we need to rely on specific laws of arithmetic or operations from arithmetic.

Exercise 10c

1. Consider the first proof in Exercises 10b.
   (a) Identify the resources that have been drawn upon in line (3).
   (b) Why does line (4) follow from the previous line?
   (c) How does line (3) fit into the proof as a whole?
2. Consider the second proof in Exercise 10b.
   (a) Identify the resource that has been drawn upon in line (2).
   (b) Why does line (2) follow from the previous line?
   (c) How does line (3) relate to an intuitive overview of the proof?
3. Consider the following theorems. Make sure that you understand each theorem. Further, find a proof of each theorem. Describe the logical structure of the proof, provide an intuitive overview of the proof and add a comment to each line of the proof.
   (a) The sine rule.
   (b) A theorem whose proof you are struggling to understand.

▶ Extension material: $\sqrt{2}$ is not a rational number

So far in this chapter we have only considered the relatively simple direct form of proof. However, in this case-study we will apply the principles developed earlier in the chapter to help us make sense of a further form of proof, *proof by contradiction*. In particular, we will consider a proof of the well known theorem that the square root of 2 is not a rational number, or is irrational. We start by providing a commentary on the theorem, and this is followed by a formal statement of the proof and then a commentary on the proof.

Commentary on the theorem

In seeking to understand this theorem, our strategy is to employ at least some of the skills from Part I of the book. If we analyse the theorem we can see it will help if we have a clear idea of what a rational number is. The definition of a rational number may be stated as follows:

> A number r is rational if there exist integers p and q, with no common factors and $q \neq 0$, such that $r = p/q$.

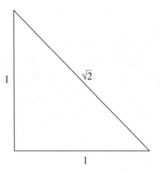

FIGURE 10.3 √2 AS THE LENGTH OF A HYPOTENUSE

Or, in more informal terms, we can say that all rational numbers can be expressed as a ratio of two integers, and thus an irrational number cannot be expressed as a ratio of two integers. We have already seen examples of both rational and irrational numbers in Chapter 2. Finally, it also helps to provide a visual representation of the number $\sqrt{2}$. The classic example in this context is to point out that $\sqrt{2}$ is the length of the hypotenuse in the right-angled isoceles triangle given in Figure 10.3, or, in effect, the length of the diagonal of a unit square. The theorem thus states that the length of this hypotenuse cannot be expressed as a ratio of two integers. Ancient Greek mathematicians were shocked when they first realised that this was the case.

What is this theorem useful for? In many ways this question boils down to the question of what irrational numbers are useful for, because this theorem demonstrates the existence of one of the infinitely many irrational numbers. A large proportion of all mathematics in fact makes use of irrational numbers, from arithmetic to integration and differentiation, and a huge range of applications of mathematics thus depend on the use of irrational numbers.

Proof that $\sqrt{2}$ is not rational

We assume $\sqrt{2}$ is rational. (1)

Hence $\sqrt{2} = p/q$, where integers p and q share no common factors and $q \neq 0$ (2)

$\Rightarrow 2 = p^2/q^2$ (3)

$\Rightarrow 2q^2 = p^2$ (4)
$\Rightarrow p^2$ is even (5)
$\Rightarrow p$ is even (6)
$\Rightarrow p = 2a$, where a is an integer (7)
$\Rightarrow 2q^2 = 4a^2$ (8)
$\Rightarrow q^2 = 2a^2$ (9)
$\Rightarrow q$ is even. (10)
By the above, both p and q share the factor 2. (11)

Statements (2) and (11) contradict each other. (12)
Hence our assumption in (1) was false. (13)
Hence $\sqrt{2}$ is not rational. (14)

Commentary on the proof

We are now in a position to try to make sense of the above proof by focusing in turn on the overall structure of the proof and then on the various details of the proof. The form of proof which has been employed here is *proof by contradiction*. We assume that a statement is true, but then find that a contradiction logically follows from this assumption. That is, we find that a statement is both true and false at the same time which, as we have already seen in Chapter 6, is not allowed. The only acceptable conclusion is that the original statement we assumed to be true must in fact have been false. We summarise this logical structure in Figure 10.4. In terms of a more intuitive overview, the proof in this particular example uses statements to the effect that several numbers are all even in order to set up a contradiction.

As for some of the details, we need to see how each line of the proof relates to the overall structure of the proof. Thus lines (2)–(11) are all about setting up the contradiction that flows from line (1). Line (12) then points out the contradiction and this leads in lines (13) and (14) to the conclusion that $\sqrt{2}$ is irrational. In addition, it is important to be able to see why each line is justified in itself. For instance, line (2) follows from line (1) by drawing on the definition we gave above of a rational number. And if we were initially to chose integers p and q, which did share a common factor, then we can legitimately cancel these common factors to give integers which would not share any common factors. It takes more work to see why

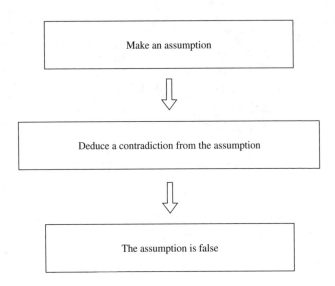

FIGURE 10.4 PROOF BY CONTRADICTION

line (5) follows from line (4), but it can be done and this is left as an exercise for you.

Extension Exercise 10

1. Employ the skills of Part I to ensure that you fully understand each of the following theorems:
   (a) The Fundamental Theorem of Algebra.
   (b) The Central-Limit Theorem.
2. Consider the following theorems. Make sure that you understand each theorem. Further, find a proof of each theorem. Describe the logical structure of the proof, provide an intuitive overview of the proof and add a comment to each line of the proof.
   (a) The Prime Factorisation Theorem.
   (b) Consider a real-valued function, f. Then f is bijective $\Leftrightarrow$ there exists an inverse function for f.

> **Reflection**
> How effectively did you tackle the Exercises in this chapter? Can you think of any ways to improve how you tackle the Exercises in the rest of this book?

Summary

- Make sure that you understand the theorem.
- Describe the logical structure of the proof and provide an intuitive overview of the proof.
- Add a commentary to each line of the proof.

Part III
Study

11 Studying Actively

> **This chapter aims to**:
>
> help you to meet the practical challenges of your own study of mathematics and its applications.

▶ Introduction

One minute you may be listening to your tutor explaining a new concept, later you will be working on the solution of a problem, and then you might have an opportunity to work through a set of problems alongside a group of fellow students. Clearly there are a variety of ways to study mathematics and its applications, and the particular approach that you follow will in many ways be determined by your particular institution or tutors.

However, it is important that you do not just let your tutors set the agenda for your study: you need to take the initiative if you are to succeed. The rest of this chapter is designed to help you to do this. Chapter 12 then focuses more specifically on how you can take the initiative when using technology to help you complete mathematical tasks.

▶ Study involving tutors

There is indeed a temptation for you to let your tutors set the agenda when you are in direct contact with them. After all, they are experts. Unfortunately, your study can easily become a process of recording what the expert has to say and then trying to reproduce their words in an examination or in coursework. But the ability to reproduce a list of meaningless symbols on a page is of little use when tackling a new task. For instance, it is one thing to be able to solve a particular type of differential equation; it is quite another thing to know the type of situation in which you should make use of it. When interacting with your tutors it is therefore essential that you actively try to develop your own understanding rather than seek to let the tutor do the thinking for you.

> Take the initiative in your contacts with your tutors.

Lectures

The classic example of a tutor taking centre stage is the lecture: the tutor talks to a group students on a particular topic, often with the aid of chalk and a blackboard. Now you might think it would be difficult for you to be active in the context of a lecture, but as we shall see this is not the case.

Anecdotal evidence suggests that once a typical mathematics lecture gets under way most students find it hard to follow the lecture. Ideas are often thrown at you far too quickly for you to understand them straight away, and for many students this is very frustrating. In one way it is possible to cope with this situation simply by adjusting your expectations. You need to accept that many ideas in mathematics and its applications will take time to comprehend fully, and we will see very shortly that it is important to take time after a lecture to make sure that you really have taken in the ideas presented in the lecture.

It is also true, however, that in order to get the most out of a lecture you need to make sure that you are adequately *prepared* for it. Preparation is needed so that you stand a chance of following the tutor's explanations, and following their explanations during the lecture is the first way in which you need to be active during a lecture. This might seem obvious advice, but it is surprising how many students wander into their lectures without doing any preparation at all. After all, how many students would expect to follow a discussion on a work of literature without having read the book?

Ideally the preparation you do before a lecture needs to be undertaken on three levels, which we can outline as follows:

1. Try to make sure as far as possible that you understand the earlier material in the course. In doing this, you will find it useful to employ the skills of Part I. Go through your existing course notes and provide extra examples, draw additional visual images, analyse the key ideas and make sure that you are fluent in your use of all the symbols.
2. A lecture will normally contain several new ideas and, as we saw in Chapter 5, each of these ideas will depend on other, more basic, ideas. So if at all possible find out in advance which new ideas will be contained in the lecture, and then make sure that you understand all of the more basic ideas on which they depend.
3. Read ahead in the relevant textbook about the ideas to be covered in the next lecture. There is of course no need to have understood everything in advance, but it will help if you have made an attempt.

Before we take a look at how you can be active both during and after the lecture, it is worth pointing out a further way in which many readers will be able to prepare more effectively for their lectures. We have seen that ideally you should understand the earlier material in a course, but you should also understand other relevant basic material. In particular, you should thoroughly understand the area of mathematics called elementary algebra. You need to be able to complete algebraic problems mentally, without resort to pen and paper. After all, most tutors will not take time in a lecture to explain algebraic manipulations for you: they will expect you to understand them quickly and without assistance, and if you regularly spend time in your lectures puzzling over algebra then you should not be surprised if you find the lecture as a whole hard to follow.

> The appendix provides advice on how to gain the level of algebraic fluency you are likely to need. Take a look at the appendix now to make sure that you have reached the appropriate level.

As far as the *lectures themselves* are concerned you need to make sure that you take a good set of notes and take as much part in the lecture as possible.

Simply copying down what is written on a blackboard, for instance, is not enough. This will not give you a comprehensive set of notes. The tutor may make additional comments that are not written up and which could well provide the insight needed to make sense of the lecture. You should also try to record some of these comments as well and, if you do not have sufficient time to note down these additional comments during the lecture itself, you can at least make a mark on the notes to remind you to add in the additional comments later.

Furthermore, some tutors try to interact with their students during their sessions. You will learn more if you try to answer their questions. If there is scope for you to ask questions as well, then do so. Otherwise note down any questions that occur to you and then find out the answers later on.

Finally, you need to take the initiative *after the lecture*. Most lectures are packed full of new ideas. It therefore becomes essential to annotate your lecture notes using the skills covered earlier in the book. Add in your own examples, particular instances, visual images, translations of symbols, analysis of key ideas and connections to other relevant ideas. Identify the logic that has been employed. It will help if you allow space when writing your lecture notes to add in these annotations. Provide an overview of each part of your notes, and then point out how different lines relate to that overview. Annotating your notes helps you to make sense of the ideas covered in the lecture and ensures that you take the initiative in your own learning. You are then much more likely to remember what you have learnt.

Get the most out of your lectures by:

- preparing adequately for them;
- participating in the lecture;
- annotating your lecture notes.

Tutorials

Tutorials are sessions in which you have an opportunity to ask your tutor questions. Again, you are more likely to gain if you are sufficiently well prepared for these sessions. This could involve bringing with you a list of questions for your tutor, but try also to use the opportunity to improve your grasp of the skills covered in this book. For instance, you could ask your tutor to provide an overview of a proof you have been struggling to understand, or you could take along a relevant visual image and ask your tutor to point out how it relates to an idea you have been struggling to comprehend. You may have found it difficult to provide a full range of examples of a certain concept; if so, get your tutor to fill out your collection of examples for you.

Do not just focus in these sessions on your assignments, something that is only too easy to do. You may well also have an opportunity to ask questions on the content of any lectures. Doing this again helps you to take the initiative in your study, rather than simply relying on the tutor to set the agenda for you.

> Get the most out of your tutorials by:
>
> - preparing questions for your tutor;
> - focusing on study skills.

Examples classes

Students usually have an opportunity to work through a series of problems with a tutor on hand for support. In some settings this will be called an 'examples class'. A common way to make use of these classes is for a student to work through a problem until he or she becomes stuck, and then to ask a tutor for help about what to do next. The tutor might then provide a full, or perhaps partial, solution for the student. This approach, however, is limited in its usefulness because it encourages the student to rely on the tutor.

A more helpful approach is to concentrate on developing your problem-solving skills. Employ the skills of Part I of the book in trying to understand the problem. And if you find this challenging on any given problem then ask your tutor to help you employ these skills, or say to the tutor something like the following: 'I am stuck at this point here. If you were in my place, how would you try to overcome this block?' Alternatively, spell out your plan of attack on the problem to the tutor or to one of your peers, or ask someone to summarise the approach that you took in trying to solve the problem. Taking this kind of attitude to your work will not only help you reach solutions to problems: it will also help you to develop your problem-solving ability.

> Get the most out of your examples classes by:
>
> asking questions that focus on problem-solving strategies.

Exercise 11a

1. Annotate your last set of lecture notes using the skills covered in Part I of the book.
2. Prepare a list of questions for one of your tutors to answer, and then ask your tutor to answer them!
3. Devise a set of questions that you would be willing to ask your tutor on how to solve a particular problem.

▶ Studying with other students

Students are increasingly being expected to study alongside one another in small groups. This partly reflects the increasing importance being given to the development of communication skills, but it also recognises that students have a great deal to gain from working with one another. So try to get the most out of your work with other students.

1. Be clear about what you are trying to achieve in your group work. It may help if you agree an agenda at the start, to keep you on track.
2. Pay attention to the *tasks* that you are asked to carry out when working in small groups with other students. Make sure that you do write the report, solve the problem or prepare the presentation.
3. Also pay attention to the *social side* of the group. Get to know one another. Make sure that everyone is involved in the work of the group. Go out for a social activity together. Your group will operate far more effectively if you get on with one another.
4. Develop together your grasp of the skills and strategies covered in this book. Encourage one another to look for examples and visual images. Compare annotations of your notes with someone else's annotations.
5. Be prepared to ask one another questions. Explaining to someone else what you think a problem involves is an excellent way of clarifying your own understanding of the problem.
6. Group-work in mathematics and its applications carries its own challenges. Different members of the group are likely to learn at different rates, so be patient with one another.
7. If your group does not work out well, be ready to ask your tutor's advice. The earlier you do this the easier it will be for them to sort out any problems that may have arisen.

Exercise 11b

1. Think of some group-work in which you have recently engaged. Describe both how you tackled the task and how the group got on with one another.
2. Choose a problem that you are struggling to solve. Explain to a friend how you have been trying to solve it.

▶ Independent study

However much time you spend studying in the company of other people, you will also need to study on your own. After all, a tutor or fellow student cannot give you understanding, so it is worth looking here at two of the main ways in which you are likely to study on your own.

Reading texts

You will usually be expected to read a range of texts, whether lecture notes, textbooks, handouts or material on the Internet. But not any kind of reading will do; you need to make sure that you read actively.

We saw in the section on lectures that it is worth annotating your lecture notes by making use of the skills covered in Part I. The same holds true for any text that you have to read. So again, add in your own examples, particular instances of examples, analysis of key ideas, definitions of symbols, connections to other ideas and so on. Most mathematical texts are densely packed with ideas, so if you fail to annotate a text you are unlikely to fully understand it.

You might also find it helpful to read a text on two levels, just as we did when seeking to understand proofs of theorems. First of all gain an overview of the material, and then focus on the individual details of the text. Spell out exactly how the details relate to you overview of the text. Changing the focus of your attention in this way gives you a different perspective on the material.

The aim in all of this is to avoid reading a text five or more times, struggling to understand how one line follows from the last. This passive approach would just encourage you to try and reproduce the text in your mind. Taking an active approach to reading is far more likely to result in a genuine understanding.

> Get the most out of your reading by:
>
> - annotating the texts you read using the skills of Part I;
> - gaining an overview of the text and then focusing on the details.

Practice

In order to achieve mastery at anything you need to practise. This is as true of mathematics and its applications as it is of basket-weaving or conducting an orchestra, but it is not any kind of practice that will do. Many students end up mindlessly solving one type of problem again and again until the solution is memorised. Unfortunately this approach will not enable you to solve problems that are different from your model problem, and as you study mathematics at higher levels you are increasingly expected to be able to solve non-standard problems.

Instead, you should aim to practise a variety of tasks (the more varied the better). For instance, when trying to master the solution of a certain type of differential equation, you might want to solve problems that require you to explore a variety of real-world situations that are modelled by the relevant type of differential equation, or you might want to tackle problems that require you to do more than just find the solution of a differential equation.

The range of practice you require will vary according to your area of study, but you will always want to introduce variety into your practice. Just as we sought to avoid re-reading a text again and again, the advice here is to avoid tackling virtually the same kind of task again and again. This only encourages a passive approach to learning.

> Find ways of introducing some variety into your practice of mathematics and its applications.

Exercise 11c

1. Annotate part of a text using the skills of Part I of this book.
2. Choose a short piece of text. First of all summarise it, and then make sure that you understand every detail of the text.

3. Take some material you are currently studying. How can you increase the variety of ways in which you practise the main tasks connected with this material?

> **Reflection**
> What practical steps will you take to improve the ways in which you study mathematics and its applications?

Summary

- Take the initiative in your study. Do not just let your tutors set the agenda.
- Make use of the skills covered in Part I of the book to ensure you fully understand the ideas with which you are dealing.

12 Using Technology

> This chapter aims to:
>
> help you use technology effectively in completing tasks from mathematics and its applications.

▶ Introduction

Technology can now help you to carry out a wide range of mathematical tasks. Computer algrebra systems such as *Mathcad*, *Mathematica* and *Derive* allow for the completion of numerical calculations, the manipulation of symbols and the production of visual images. Some hand-held graphical calculators afford similar capabilities. Programming languages allow for the completion of an even wider range of tasks. Then there are geometrical software packages that enable the manipulation of geometrical objects and interactive graphical tools for producing and working with visual images. There are spreadsheets and software that enable you to calculate relevant statistics.

Such software packages or technology usually accept a task or input and return a solution or output, although the nature of both the input and the output will vary. Thus a computer program or a spreadsheet might take as input some information that would enable it to evaluate a definite integral. It would return as output an approximation to the value of the integral. Or a computer algebra system (CAS) might accept an algebraic expression as input and return as output an equivalent algebraic expression represented as a product of factors. In this case no approximations would be made in producing the output.

As a consequence of this new technology, the practice of doing and applying mathematics is changing. Mathematicians and those who apply mathematics need no longer carry out for themselves many tedious calculations, and graphs that would take hours to draw by hand can now be produced in a matter of seconds.

Even allowing for this, however, there is no guarantee that a tool such as a CAS will be of much use to you. While it is true you may save time if you get a computer to carry out a task for you, simply carrying out the task is often the least of your worries. You may well be more worried about which task you should carry out in the first place or about what the output produced by the computer actually means.

This chapter is therefore designed to ensure that your own study of mathematics and its applications benefits as much as possible from the assistance provided by technology.

However, before we consider how you can make effective use of technology, there is an important point to be made. In order to use a new software package appropriately, you need to know how to operate it, so spend some time initially learning how to operate the technology. It can be tempting to try and muddle your way through, but if you do so you are likely to waste more time trying to implement the software than you would have taken mastering it in the first place.

> So do complete the online tutorial when you first encounter a new software package. Do attend the training sessions that are offered to you.

▶ Use the technology

Once you have mastered a particular technology or software package, you need actually to make use of it and, in particular, you need to take the initiative. Do not just be limited by the immediate assignment that you have to complete or by a specific task your tutor has asked you to address. Technology provides you with an immense toolbox: use it to support your study more widely.

If your use of technology is completely separate from the rest of your work then you should not be surprised to find the technology of little use. Ideally you should obtain a personal copy of the relevant technology. This is the best way to make your use of technology as natural to you as your use of pen and paper. After all, imagine doing mathematics without any recourse at all to pen and paper. This is surely unthinkable. And you can benefit from technology just as you can benefit from pen and paper.

It is also true that you need to be clear about what you want to achieve when you experiment with the technology. If you simply play with a piece of software in an unfocused manner then learning is unlikely to result.

> The recommendation here is that you use technology to help employ the skills of Part I and to carry out the tasks of Part II.

In this chapter we will look at how we can actually ensure that technology enables us to use examples, think visually, cope with symbols, think logically and make connections. Employing technology in this fashion will help ensure that you use it to help you make connections between ideas, which as we have seen provides a highly effective way of studying mathematics and its applications. We will also see how technology can help us take advantage of the features that characterise the problem-solving of the expert, focus on all the stages of the modelling cycle, and make sense of proofs of theorems.

In using technology to help us with these skills and tasks, it is worth recalling the nature of the technology we are dealing with, because its nature will determine how to make use of it. In this case, the technology we are concerned with usually involves entering an input and the computer, say, returning an output. The focus in employing these skills and in completing these tasks thus needs to be on the choice of input, on making sense of what happens to that input and on interpreting the output.

▶ Employ the skills

So how, then, can we actually make use of technology to help us employ the skills that were outlined in Part I of this book? The easiest way to see this is to consider each relevant skill in turn.

Using examples

One of the primary uses of the technology we are concerned with is to undertake calculations. Following the pattern established in Chapter 2, it is important to gain a feel for a given type of calculation by working with a varied set of examples of the calculation, so carry out several different versions rather than just one calculation. Change the input slightly and see what happens, or carry out a more unusual version of the calculation. You can then ask whether the output from the new calculations differ as you would expect from the initial calculation, and if there is a difference that surprises you, can you account for it?

For instance, if you are evaluating a definite integral using a CAS you could change the bounds of integration and see how the output is affected. Does the area under the curve change as you would expect? Or if you are calculating a statistic, such as the standard deviation or a χ^2, then make use of more than one set of data. Calculate the same statistic for completely different sets of data. You can then see how the statistic changes as the data changes and this will give you a sense of the way in which any given statistic describes the data.

You may also be able to use both the input and the output for a calculation as a springboard for considering particular instances of examples of ideas. Perhaps a CAS returns a specific function as the output. This provides an excellent opportunity for you to consider particular instances of the function.

- Carry out a variety of calculations.
- Use both the input and the output as a prompt to consider particular instances of examples.

Thinking visually

One of the most powerful uses of technology is to visualise abstract mathematics. As we have seen, visual images can be of great assistance in understanding concepts, theorems, problems and applications, so take full advantage of your graphical calculator, your CAS or that specialist graphics package to produce images of your own choosing. Ideally you should get used to creating a whole collection of visual images for each topic you are working on, assuming of course that particular area of mathematics and its applications allows scope for this.

It is, of course, important to do more than just produce visual images. You should also employ the skills outlined in Chapter 3 to make sense of your images. Both pick out some of the details of the image and try to gain an overall perspective of the image, and then link these observations to other mathematical results or to the real world. After all, a piece of software may draw a visual image for you, but that does not guarantee you will learn anything as a result.

- Produce visual images as a matter of habit.
- But make sure that you look at your images flexibly and connect them with other mathematical ideas or with the real world.

Coping with symbols

Technology primarily confronts you with a whole new set of symbols to learn before it can begin to help you to cope with all the rest of the symbols that you face in your study. It is therefore worth applying the skills we developed in Chapter 4 to make sure that you are fluent in your use of these symbols. Do you really understand all of the symbols that are present in both your input and your output? Can you provide a full and instant translation of what they actually mean?

However, if the symbols used by the technology are simply different but corresponding versions of more standard symbols you will also benefit if you can explicitly spell out exactly how such symbols are linked to each other. This will help you make more effective sense of both the input and the output, and in addition you will enrich your appreciation of the more standard mathematical symbols.

Make sure you know the full meanings of the symbols contained in your input and output.

Thinking logically

Some software will be able to carry out various logical tasks for you. Making use of these capabilities will enable you to improve your understanding of logic, particularly if you are able to connect the operations carried out by the technology with the formal language of logic.

However, technology can also become a springboard for you to examine the logic that underlies a piece of mathematics or an application of mathematics. Certainly with a CAS performing symbolic calculations for you, it will be important for you at least occasionally to carry out the calculation for yourself so that you can make explicit the exact logical way in which the output is related to the input. For instance, it might be the case that the output simply represents an equivalent statement to the input, or it might be the case that the input and output are not related via a logical relationship but are simply different representations of the same mathematical object.

Making connections

Perhaps the most obvious connections to look for when using technology are connections between the input and the output. One way to do this is of course to recreate using pen and paper the actual process (or at least a version of this process) that the technology used to transform the input into the output, and you can also gain a feel for the way in which the output depends on the input by varying the input slightly, as we have already seen.

It is also important, however, to link the input and output with other ideas from mathematics and its applications. Take your course notes and compare them with the symbols on the screen. Juxtapose them against the input or the output, and spell out any relevant connections that you see between the screen and the notes.

In general it will help if you are aware of the type of connection you are looking for. The usual suspects will all be relevant, from the symbols on the screen all being particular instances of some more general result, to the symbols representing examples of concepts to the real world, and so on.

> Think of the technology as a window onto mathematical meaning or onto the real world. To focus only on a computer screen itself would be to ignore all of this wider meaning.

Exercise 12a

1. Choose a particular mathematical task that you are performing on a computer.

(a) Relate the contents of the screen to other mathematical ideas.
(b) Vary the input slightly. Does the output change as you would expect?
(c) Spell out the full meaning of both the input and the output.
2. Choose a visual image that you have created on a computer or graphical calculator. Link both some of the details and an overview of the image to other mathematical ideas.

▶ Carry out the tasks

It is also important to ensure that you use technology to help in solving problems, modelling the real world and constructing proofs, as developed in Part II of the book. We can take each of the three main tasks in turn and see how technology can help us complete them.

Solving problems

Technology can obviously carry out certain tasks for you, and this is of course very useful in problem-solving. For example, if you are asked to solve a certain differential equation you might be able to get some software to solve it for you. Indeed, the range of problems which technology can now solve is truly staggering, and you will certainly want to employ technology to help you solve problems as a matter of course.

However, we can also observe that when a piece of software is able to carry out tedious calculations for you, then time will be freed for you to concentrate on understanding the problem, devising plans and reviewing the situation. Furthermore, we have already seen that technology will be of assistance in employing the skills of Part I of the book, and since all of these skills are relevant to gaining an understanding of the problem, as we saw in Chapter 8, they are thus all relevant to solving problems. Finally, by checking both the input and the output, to make sure the relevant task is being correctly performed, you gain an opportunity to review your solution of the problem.

Applying mathematics

The major way in which technology is useful for applying mathematics is in helping us to solve problems that stem from models of the real world. For those readers whose study includes applications of mathematics, this will be a key way in which you use technology.

So you will need to be aware of the different ways in which technology might complete solutions of problems for you. For instance, you may have evaluated a definite integral approximately, say by software employing numerical analysis, or you might have evaluated it exactly, say by a CAS. But if only an approximate answer is returned by the software you will need an awareness of this to inform the conclusions that you draw from the model.

> In general it is important that you retain an understanding of how the technology actually carries out the task for you.

For one thing, this helps to give you an appreciation of the tasks technology can carry out and thus of what input is acceptable. But this will also be of use in helping you to make sense of the output. It might be that the input has been incorrectly entered, and it is only because you are aware of what the process carried out by the computer ought to lead to that you can pick up on the incorrect input.

Just as technology could support elements of problem-solving beyond merely carrying out a plan of attack on the problem, so technology is useful in stages of the modelling cycle other than just the solution stage.

For instance, it is also important to use the input and the output as a springboard to interpret the mathematical ideas involved. Can you match the real-world examples to the mathematical examples provided by the software? Can you identify the real-world concepts of which these are examples? How are the relationships between concepts from the real world expressed in terms of relationships between the mathematical ideas? This kind of interpretation is particularly relevant to the third, fourth and sixth stages of the modelling cycle.

Such interpretation further provides an important check to see whether the input and output are both free of errors. If you interpret the output and find that your interpretation is meaningless then you will need to review the input; but if your interpretation of both the input and output seems reasonable, then you gain extra confidence that your model is appropriate.

We can, finally, point out that one way of validating a model is to see how robust the model is. You might, for instance, want to test what would happen if you made a few small changes in the parameters of the model. Does the model still predict a reasonable outcome? If you had to repeat a long calculation by hand you would be much less likely to make these tests, but with the appropriate technology at your disposal such testing is straightforward.

Constructing proofs

Technology will be of use in helping you to understand and construct proofs of mathematical results. Software is available that actually validates theorems of certain types, although in practice you are probably less likely to exploit this kind of capability at this stage of your studies.

However, it might well be the case that technology will help you in completing a part or even the whole of a proof, the validity of which is left to you to determine. For instance, a CAS may carry out for you a task that is relevant to the proof. You might start with an established result and ask a CAS to make a substitution into this result, with the CAS returning as output the theorem you are seeking to prove.

Of course, you will also want to use technology to help you employ the skills in Part I of this book. You should never underestimate the power, for instance, of a visual image in helping you to make sense of a proof. The choice is always yours as to whether you make use of technology in developing your understanding.

▶ Case-study: use of a computer algebra system in modelling population growth

In conclusion to this chapter, it will be worth taking a look in greater detail at a practical example of using technology. More specifically, how might we take the initiative in using a CAS to understand the solution to a model of population growth?

Now suppose a population exists for which the size of a population at time $t > 0$, $P(t)$, is determined by the following differential equation, where a is a parameter:

$$\frac{dP}{dt} = aP \cos t$$

How can we make use of a CAS in order to help us understand this model of population growth?

The most obvious use to make of the CAS is of course to ask it to solve the differential equation. If we carry this out, we find that for a constant c,

$$P(t) = ce^{a \sin t}$$

At this point the reader is encouraged to consider the effect of different values of both a and c and to provide particular instances of the function P, along with matching instances from the real world.

Of course, it is also possible to use the CAS to help us visualise what is going on in this situation. We can plot graphs of the solution for specific values of a and c. For instance, taking $a = 2$ and $c = 3$ we obtain the graph given in Figure 12.1. You should be able to match details of this graph to particular instances of the function $P(t)$ for these values of a and c. As far as an overview of the image is concerned, we can see that the population oscillates as time increases. Try plotting graphs for other values of a and c to develop a feel for the more general version of the solution.

However, it is still hard to see how this solution and its graph relate to the original differential equation. It is therefore worth exploring these relationships in greater detail before we make further use of the CAS. In particular, the differential equation summarises how the rate of change of the size of the population varies with time. Alternatively, we can say that the differential equation

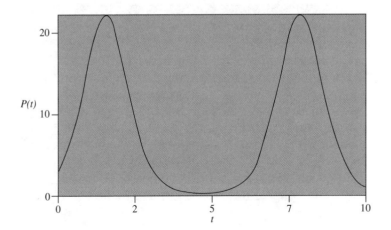

FIGURE 12.1 THE GRAPH OF $P(t) = 3e^{2 \sin t}$

details how steeply the population is increasing or decreasing at any given time. A function $P(t)$ will be a solution of the differential equation if the slope of the function at each point t is as predicted by the differential equation.

We can now use the CAS, if we take for instance $a = 2$, to visualise the value of dP/dt for different values of t and $P(t)$: that is, at different points in the Cartesian plane $(t, P(t))$. In particular, we can represent each of these gradients dP/dt by a line in the direction specified by the gradient (for our purposes the length of each line is not important). This is effected in Figure 12.2 by plotting vectors of the form:

$$\begin{pmatrix} 1 \\ 2P \cos t \end{pmatrix}$$

You might find it helpful to confirm for yourself that the direction of each vector of this form matches the corresponding value of dP/dt.

If we now superimpose the graph of one of our solutions to the differential equation on to this vector field, we can appreciate in Figure 12.3 that the slope of each tangent to the graph of our solution is indeed as predicted by the differential equation. You can see the way in which changes in the gradient of the tangents are mirrored by changes in the shape of the graph of the solution.

If we superimpose other valid solutions of the differential equation, as is done in Figure 12.4, we can also appreciate that an infinite number of curves all provide tangents with slopes that satisfy the

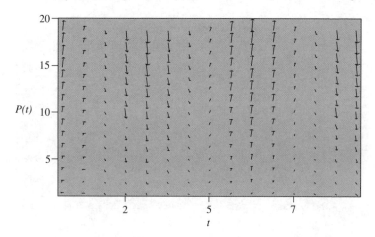

FIGURE 12.2 VECTORS REPRESENTING SELECTED VALUES OF dP/dt AT DIFFERENT POINTS OF THE PLANE $(t, P(t))$, WHEN $a = 2$

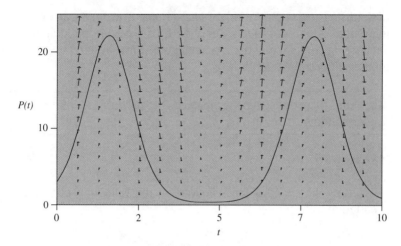

FIGURE 12.3 A GRAPH OF THE FUNCTION $P(t)$, WHEN $a = 2$ AND $c = 3$, SUPERIMPOSED UPON VECTORS REPRESENTING SELECTED VALUES OF dP/dt AT DIFFERENT POINTS OF THE PLANE $(t, P(t))$, WHEN $a = 2$

differential equation equally well, and thus we have an infinite number of solutions to the differential equation.

With the aid of a CAS we have now reached a much fuller understanding of the solution of our model of population growth, and it would of course also be possible to explore other aspects of this

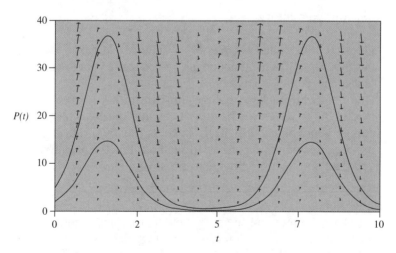

FIGURE 12.4 GRAPHS OF THE FUNCTIONS $P(t)$, WHEN $a = 2$, $c = 5$ AND WHEN $a = 2$, $c = 2$, SUPERIMPOSED UPON VECTORS REPRESENTING SELECTED VALUES OF dP/dt AT DIFFERENT POINTS OF THE PLANE $(t, P(t))$, WHEN $a = 2$

model using the CAS. We might wish to see how well our existing differential equation models the growth of a certain population by considering some actual data, or we might want similarly to explore another differential equation.

> Technology needs to be employed as part of a search for understanding.

Exercise 12b

1. Explore using a CAS the solutions of the following differential equation:

$$\frac{dy}{dx} = \cos x \, e^x$$

2. Complete your next relevant coursework assignment without any (or with as little as possible) assistance from technology. Then, for the following assignment, make as extensive use as possible of technology, following suggestions given in this chapter. Can you both describe and account for any differences between the work you produced for these two assignments?

Extension Exercise 12

1. Return to the extension material for Chapter 8 and using technology plot a graph of the function g. Can you relate your image to the text of the extension material?
2. Find a software package that you think will be of use to you in your studies but to which you have not been introduced as part of your course. Learn how to use the package on your own, making use of printed documentation or help available within the package itself, and then employ the software to assist your studies.

> **Reflection**
> Consider for a moment the ever-growing role that technology is playing in our society. By using technology more widely in your studies you will also be acquiring a valuable skill for your future career.

Summary
Use technology to:

- assist you in a search for understanding;
- employ the skills of Part I and carry out the tasks of Part II.

13 Succeeding in Assessment

> **This chapter aims to:**
>
> help you to meet the practical challenges of assessment in mathematics and its applications.

▶ Focus on communication

You may be able to solve a problem or understand a theorem, but unless you can communicate your work effectively to your tutor or examiner then you will not do well in assessment. The quality of your communication affects how well you succeed in your study, so this chapter is designed to ensure that your communication is rated as highly as possible by your tutor or examiner.

In particular, we will look at some specific characteristics of effective communication: your communication needs to be fit for the purpose, clear and polished (see Figure 13.1). We will see later in the chapter the different ways in which these principles are applied in the specific kinds of assessment you are likely to face.

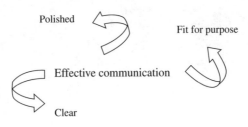

FIGURE 13.1 CHARACTERISTICS OF EFFECTIVE COMMUNICATION

Is it fit for purpose?
This question addresses whether or not your communication is what your tutor or examiner is looking for. In the context of mathematics and its applications you need to make sure your work is *coherent* and *accurate*.

1. Coherence primarily comes from a clear structure. This reflects the understanding that mathematics and its applications is an integrated system of thought rather than a disparate collection of facts, but you also need to make it clear how each detail of your communication relates to the structure.
2. Accuracy is of course also important in mathematics and its applications. Symbols need to be used in accord with their precise meaning. Mathematical ideas and concepts from the real world are characterised by formal definitions. If you ignore these definitions then your work will be bedevilled by errors and inaccuracies.

Is it clear?
Have you ever been in a situation where you have said something and your listener has misunderstood what you meant? It is not enough for you to understand what you mean when you write or say something; you also have to communicate your understanding to your audience. To put it another way, your tutor can only assess what you have actually communicated and not what you meant to communicate. This means that it is essential that your communication is *clear*: your work must not leave itself open to being taken to mean something completely different from what you intended.

Is it polished?
Imagine an actress trying to hold an audience while mumbling half-forgotten lines. Clearly for a performance to be effective it needs to be *polished*, and the same is true when you are trying to communicate mathematical ideas.

1. In work submitted during your course you usually have an opportunity to revise your writing before you submit it. Take advantage of this opportunity. It is, after all, unrealistic to expect that you can turn out a piece of work that is coherent, accurate and clear at the first attempt. You need to ask whether there are any ambiguities, inaccuracies or errors in your work; you need to make sure that

your work has a clear structure; and you need to make sure that someone else can make sense of what you have written.
2. In situations where there is little time to revise your communication, such as during an examination or when giving a presentation, you need to make sure that the revision has gone on beforehand.

▶ Coursework

Students are increasingly being required to complete work that will be assessed during their courses, and this includes not only the more familiar solution of problems but also projects, reports, essays and presentations. It is therefore worth looking in greater detail at how to ensure that you can communicate your understanding effectively in such contexts.

Mathematical text

The most common type of assessment you will face is writing pieces of mathematical text. You might be describing the solution of a problem or the creation of a mathematical model.

So how can you ensure your text is *coherent*? Texts that only involve mathematical ideas primarily derive their coherence from logic. (Logic was discussed in Chapters 6 and 10.) You need to make sure that a logical argument is apparent in your text. Compare, for example, the following two short pieces of text:

$$3x + 2 = x, \quad x = -1$$
$$3x + 2 = x \Leftrightarrow x = -1$$

In the first line the logical connection between the two equations is not at all evident, whereas in the second line it is clear that these two equations are equivalent to each other. The first of these lines is somewhat incoherent; the second is clearly more coherent. Or alternatively, consider the following text:

$$\int \frac{1}{x} \log_e x \, dx =$$
$$t = \log_e x \quad \text{and} \quad dt = \frac{dx}{x}$$
$$= \int t \, dt = \frac{1}{2} t^2 + k$$

Presumably the intention here is to evaluate the integral, but the text is somewhat incoherent. A more coherent version of this text would be as follows:

$$t = \log_e x \Rightarrow dt = \frac{dx}{x}. \quad \text{Hence} \quad \int \frac{1}{x} \log_e x \, dx = \int t \, dt = \frac{1}{2} t^2$$

Here at least the logical connection between the statements is evident.

By contrast, texts that describe an application of mathematics, while relying on mathematical ideas and thus on logic, are also concerned with the way in which the mathematical ideas actually model the real world. This of course means that you need to ensure that you point out the way in which the mathematical ideas model the real world. For example, when modelling the relationship between price and demand in Chapter 9 it was important that we pointed out the meaning of the different symbols involved in the equation

$$D = aP + b, \quad a < 0$$

It is also important to note how essential *accuracy* is to mathematics and its applications. If you make one mistake in a proof then the whole argument may well be flawed, or if you make an error in calculating the load that a particular wall can bear then a collapsed house may result. So check your text to ensure that your arguments, calculations and solutions are all accurate.

You further need to make sure that the meaning of your text is *clear*. In particular, if you refer to an object or idea then it should always be clear to what exactly you are referring. You need to be careful when using words like 'this' or 'that'. Is it genuinely clear to what you are referring? Undefined variables are also another source of confusion. Consider, for example, the equation

$$2x + 3 = 0$$

What does 'x' refer to here? If x is an unspecified integer rather than an unspecified real number then the equation will have no solutions.

For another example we can return to the text above outlining the evaluation of the integral. We improved the text by making the logical relationships between the statements evident. But even this did not really create a clear text. We can improve still further, as follows:

$$\text{Let } t = \log_e x. \quad \text{Then } \frac{dt}{dx} = \frac{1}{x}. \quad \text{Hence } dt = \frac{dx}{x}$$

$$\text{By the above } \int \frac{1}{x} \log_e x \, dx = \int t \, dt = \frac{1}{2} t^2 + k = \frac{1}{2} (\log_e x)^2 + k$$

This time at least it is clear that the variable t has been defined for the purpose of the integration, rather than just plucked out of the air for no apparent reason. Each of the steps involved in the integration has been spelled out more clearly, and we have also reached a solution which is clearly related to the expression we were originally trying to integrate, rather than a solution in terms of some unknown variable t.

In mathematical modelling it is similarly worth spelling out exactly those concepts from the real world to which your variables correspond, as we saw in greater detail in Chapter 7. For instance, if you introduce a variable which you have called s, is it clear whether it represents distance or speed? It is not enough for you to know that to which you are referring; you also need to make it clear to your reader.

Finally, you need to be willing to *polish* your text. If you only focus on the solution of a problem and not on how to communicate that solution then you will miss out on valuable marks. You may, for example, need to add extra comments or notation to the text to bring out the logical structure of your work or to point out the different stages of the modelling cycle. You need to be willing to turn a sketchy solution or argument into a coherent, accurate and clear piece of mathematical communication.

Some pointers towards writing good mathematical text:

- spell out the logic that underlies your text;
- focus on the different stages of the modelling cycle;
- be clear about that to which you are referring;
- be prepared to revise your work.

Reports and essays

Students sometimes claim that they chose to study mathematics or one of its applications because they did not like writing essays. While some courses allow you to avoid writing a single essay – as was the case with the author's own degree course in mathematics – the likelihood of any given student having to write a report or essay during a course on mathematics and its applications is certainly increasing greatly. It is therefore worth taking a look at how to ensure that the reports and essays you write are effective at communicating your meaning.

Coherence in a report or essay often comes from arguing for a particular case. For instance, if you are reporting on an investigation you carried out then you will need to argue that you carried out the investigation in an appropriate manner and that your conclusions can legitimately be drawn from your findings. The text should then be structured in a way that enables someone else to follow your argument. To start with, you will need an introduction that outlines how your argument will unfold. Then it needs to be clear how each paragraph fits into your argument. And of course it also needs to be clear how each sentence fits into the paragraph as a whole. One of the easiest ways to ensure that your paragraphs and sentences relate to the overall argument is to make liberal use of link words, such as: hence; thus; for example; also; because; so; since. Link words enable you to make apparent your structure. After all, the aim is to enable someone else to follow your writing.

Your writing of course also needs to be *accurate*, so be ready look up the meaning of any technical terms to ensure that you are using them accurately. If your text involves reporting the results of an investigation or experiment then you need to make sure that you provide a faithful representation of what actually happened, and that means that you need to keep accurate records in the first place. It would be dangerous only to rely on your memory.

Good writing is further characterised by *clarity* of reference, so write in short sentences. Long sentences are notoriously difficult to understand. In addition, be careful when using words such as 'it' and 'this', which can refer to any number of things.

Finally, it is important to make sure that you *revise* your writing. If someone else were to read the text, would they be able to understand what you meant? It can make a significant difference to your grade if you allow time to revise your writing.

> Check to see whether your writing is coherent, accurate and clear.

Oral presentations

Many courses in mathematics or its applications now require students to develop their oral presentation skills. Standing in front of your peers and explaining a piece of mathematics or how you carried out an investigation is becoming commonplace. This stems in part from the increasing value given to presentation skills; but presenting a piece of work is also an excellent test of how well you understand the material you are presenting.

In order to ensure that your presentation is *coherent*, begin by giving your audience an outline of what you will say. This makes it easier for them to follow the presentation. Then, during the presentation, it helps if you let your audience know how the points you are making relate to the outline. This is often called 'signposting'; tell your audience where they are going.

Accuracy also matters in presentations. Make sure that there are no errors in the visual aids or handouts that accompany your presentation. Take enough time to select appropriate choices of words. This is particularly important if you are to avoid reading from a pre-prepared script (something which it is very difficult to carry off well).

There are then several ways to ensure that your presentation is *clear*. You need to have a thorough understanding of the material you are presenting. Make sure that you have employed the skills developed in this book to understand the material you are explaining. It is also important to maintain a measured pace. Many students fall into the trap of rushing their presentation, but remember that you have been working with the material for a reasonable amount of time; your audience will not usually have had that advantage. Finally, your meaning will also be clearer to your audience if you allow them to become involved in the presentation in some way. Get them to provide another example of an idea you are explaining, or ask them to connect two ideas with each other or, if you are really confident, to ask you questions.

It is of course difficult to *revise* your presentation while you are actually giving it. You cannot easily say: 'No, that sentence was not as I meant it; I really meant this.' But you can polish your presentation before you give it, so practise your presentation in front of a fellow student, and get your listener to make suggestions as to how you could improve.

Feedback

Finally, as far as coursework is concerned, it is worth making a few points about feedback. As well as giving a grade, tutors will usually provide some comments or feedback on your work. This feedback might point out any mistakes you have made, indicate how you could improve your work in future and perhaps even highlight the strengths and weaknesses of your work. So how then can you get the most out of these comments?

1. Make sure that you can understand each of the comments. If you cannot see why the tutor has corrected your work, then ask for clarification.
2. Ask yourself how you could improve your work in the light of the comments.
3. If model answers have been provided to a set of problems then you can also gain feedback by comparing your work to the model answers. Carry out a point by point comparison of the ways in which your answer and the model solution differ from each other.
4. Identify changes you could make to your work to ensure it is nearer to the model sought by your tutor.

Exercise 13a

1. Which of the following two lines of text is more coherent and clear? Why?

$$2x + 7 < 1 \quad 2x < -6 \quad x < -3$$

$$\text{Consider } x \in \mathbf{R}. \quad \text{Then } 2x + 7 < 1 \Leftrightarrow 2x < -6 \Leftrightarrow x < -3$$

2. How could you make the following text clearer?

   Now $\mathbf{F} = m\mathbf{a}$. So if we take a force of 15 and a mass of 3 then $a = 5 \, \text{m/s}^2$.

3. Rewrite the following text:

$$y = a^x. \text{ So } \frac{dy}{dx} = \text{ But } \log_e y = x \log_e a \text{ and }$$

$$\frac{d(\log_e y)}{dx} = \frac{d(x \log_e a)}{dx}, \quad \log_e a = \frac{1}{y}\frac{dy}{dx}, \quad \frac{dy}{dx} = a^x \log_e a$$

▶ Examinations

Much of the advice that has been given earlier in this chapter on effective communication is also relevant to taking examinations. But examinations pose their own challenges: you are not able to consult additional resources and you only have a limited amount of time. This means that preparation for any examination is essential and that you have to use the time during the examination as efficiently as possible.

Revision
Some students seem to suffer from 'mental blocks' during examinations. They are sure they knew how to complete a question a few hours ago, and in a few hours they will know what they should have done. The best way to avoid suffering this fate is to ensure that you are fully prepared for the examination.

One of the best ways to prepare for an examination is to draw up a list of the topics that are covered by the examination. You then create a set of revision notes on each of these topics (revision notes primarily consist of a condensed version of your full set of notes). Alongside each topic in your list you should then list the types of question you might face on that topic in the examination. Finally, alongside each question you should write down a summary of how you would go about answering the question.

> Make a list of topics to revise. Each topic should be accompanied by notes, likely questions and a summary of how to answer each question.

Practice is, of course, also necessary in preparing for an examination, and the best practice is to be had on past examination papers. These enable you to practise as closely as possible for the real thing (although you will need to be careful that the syllabus has not changed). So answer previous examination papers under examination conditions (no access to textbooks and only taking the time allowed for the paper).

It is also worth noting that you not only need to practise answering questions, you also need to practise deciding on how to tackle likely questions. For instance, you can practise deciding on how to attack a problem. This ties in with the revision plan that we outlined above and will also help you to develop your problem-solving skills. It is an essential feature of the kind of practice that ensures you will avoid facing mental blocks during an examination.

> Practise on past examination papers.

The examination

Examinations occur during a set period of time in a specific location. That means you need to be in the right place at the right time with everything that you need. Examinations are also beset by regulations. You ignore these regulations at your peril. For instance, the front or first page of an examination paper will usually include some instruction which outlines what you have to do. Make sure that you follow it. There is little point answering six questions when the rubric indicates that only the first five answers will be marked.

It is usually helpful to start the examination by tackling the question that you think is easiest to answer. Getting off to a good start will help your confidence. It will also mean that you are less likely to become anxious and thus suffer from a mental block. Whatever you do, though, do not leave your best question until last.

During the examination, some students allow themselves to get caught up in a particular question and then find that they have insufficient time to complete as many questions as they need to. This danger is particularly relevant in mathematics and its applications because questions often finish with a difficult part to challenge the brightest students. The simplest way to ensure that you do not fall prey to this danger is to decide at the start how much time you will spend on each question. Never take extra time on an individual question unless you have reached the end of the paper and find that you have some spare time.

Just as it is important for your understanding of lectures that you possess a fluent mastery of algebra, it is similarly important for success in assessment. The number of marks that are lost in examinations when students make mistakes in straightforward algebraic manipulation does not bear thinking about. Make sure that you really have achieved a fluent mastery of algebra, reaching to the level outlined in the appendix.

It is finally worth pointing out that you will find it useful to employ the basic approach outlined in Chapter 8 on problem-solving. When you are answering a question, make sure you are clear about what you know and what you want to know. Do not be afraid to spend some time understanding the problem, on planning the solution and on monitoring how well the solution is proceeding. It is easy to waste time tackling the first plan that enters your head without thinking whether or not it is the best plan to carry out.

- Pay attention to the regulations and obey them.
- The time you spend on a question should be proportionate to the marks available for that question.
- Make use of your problem-solving skills.

Exercise 13b

1. Write down a list of steps you will take to improve the way you prepare for your next examination.
2. Take a topic you are currently studying. Summarise your notes on the topic. Make a list of the kinds of question that you might be asked on this topic. Summarise how you would answer each type of question.

▶ Final word

Succeeding in assessment is not a matter of luck, and neither is it simply a matter of how much time you spend studying. It is the students who effectively communicate their understanding who succeed. Putting into practice the advice given in this book will help to ensure that you are one of those students.

> **Reflection**
> Are you willing to devote time to polishing your coursework?

Summary

- Make sure that your work is coherent, accurate, clear and polished.
- Plan your revision carefully and do not let yourself get carried away in the examination.

Appendix: Mastering Algebra

> **This appendix aims to**:
>
> enable you to achieve a fluent mastery of algebra

It is one thing to be able to work laboriously through an algebraic task; it is quite another thing for the task to become so second nature to you that you can concentrate on new conceptual challenges or on solving a problem. This appendix therefore aims to enable you to complete a wide range of algebraic tasks without resorting to pen or paper.

We can outline the basic strategy that will help us to achieve this objective as follows.

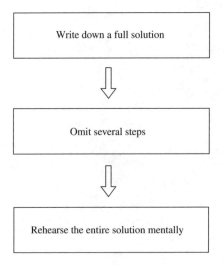

FIGURE A.1 A STRATEGY TO DEVELOP YOUR FLUENCY AT MENTAL ALGEBRA

STEP 1 The initial step in the strategy is to write down each stage of a given algebraic task.

STEP 2 The next step is to only write down one or two stages of the task. At the same time you should in turn rehearse in your mind all the intermediary stages of the task. This needs to be carried out without reference to the full written version of the solution.

STEP 3 The final step in our strategy is to put away all the other written versions of the task and simply in turn rehearse in your mind each of the stages of the task.

The strategy is also outlined in Figure A.1 and we can see it at work in Example A.1.

EXAMPLE A.1 MENTAL SOLUTION OF A LINEAR EQUATION

The task we will consider here is to solve the equation $2x + 1 = 4$.

STEP 1 We can write down a full solution as follows:

$$2x + 1 = 4 \Leftrightarrow 2x + 1 - 1 = 4 - 1$$
$$\Leftrightarrow 2x = 3 \Leftrightarrow \frac{2x}{2} = \frac{3}{2} \Leftrightarrow x = \frac{3}{2}$$

STEP 2 Start with the equation $2x + 1 = 4$. If we take away the number 1 from each side of the equation we can write down the equation

$$2x = 3$$

Finally we can then divide each side of the equation by 2 so that we give the solution as

$$x = 3/2$$

STEP 3 The final step in the strategy is to run through the whole solution mentally, so in your mind you need to subtract 1 from each side of our starting equation. Then divide the resulting equation (which you of course need to see in your mind) by 2 to yield the result that

$$x = 3/2$$

Now the algebraic task in this example might seem relatively easy to carry out mentally, even without employing the strategy we have just outlined. But the advantage is that our strategy can also be employed on more sophisticated tasks, as we can see in Example A.2. There is a real skill here that needs to be developed.

EXAMPLE A.2 MENTAL REARRANGEMENT OF AN EQUATION

The task we will consider here is to change the subject of the following equation from y to x:

$$y = \frac{2x+3}{x-5}; \quad x \neq 5$$

STEP 1 A relatively full solution is as follows:

$$y = \frac{2x+3}{x-5} \Leftrightarrow y(x-5) = \frac{2x+3}{x-5} \times (x-5)$$

$$\Leftrightarrow y(x-5) = 2x+3 \Leftrightarrow yx - 5y = 2x+3 \Leftrightarrow yx - 2x = 3 + 5y$$

$$\Leftrightarrow x(y-2) = 3 + 5y \Leftrightarrow x = \frac{3+5y}{y-2}; \quad y \neq 2$$

STEP 2 This time we only write down two intermediate stages to the task:

$$y = \frac{2x+3}{x-5} \Leftrightarrow yx - 5y = 2x + 3 \Leftrightarrow yx - 2x = 3 + 5y \Leftrightarrow x = \frac{3+5y}{y-2}; \quad y \neq 2$$

Now can you rehearse in your mind each of the stages of the task that we have omitted?

STEP 3 Finally you should be able, without reference to the full solution, to mentally rehearse in turn from memory all of the missing stages between the starting and the final equations:

$$y = \frac{2x+3}{x-5} \Leftrightarrow x = \frac{3+5y}{y-2}; \quad y \neq 2$$

If you are unable to provide these missing stages then you certainly need to develop your fluency at mental algebra.

If you carry out this strategy on a regular basis, you will soon find that you can complete a large number algebraic tasks entirely in your head. The benefits of developing such a mental facility are certainly worth the effort.

Appendix Exercises

1. Carry out the strategy outlined above for developing mental fluency upon each of the following algebraic tasks.
   (a) Solve the following equations:
      (i) $3x - 3 = x - 6$
      (ii) $3(2 - 3x) = 5x + 4$

(b) Make t the subject of each of the following equations:
  (i) $s = 2(3t - 5)$
  (ii) $p = \dfrac{3-t}{3t-6}$; $t \neq 2$

(c) Simplify the following expressions:
  (i) $\dfrac{3x - 6}{yx - 2y}$
  (ii) $\dfrac{2p^2 - 2pq}{2pq - 2q^2}$

(d) Solve the inequality $2x - 3 < 5(2 - x)$.

2. Employ the strategy outlined above on a series of algebraic tasks of increasing difficulty and of your own choosing.

Answers and Comments for Selected Exercises

▶ Chapter 2

Exercise 2a

1. The idea of a real number.
2. We need the idea of a square with sides of a given length (a unit square) and the idea of real number (which allows us to determine the number of unit squares covered).

Exercise 2b

1. (a) A non-example for the idea of the set would be the objects 1, 5, 8, 11. We have four separate objects: they have not yet been collected together! As a further exercise, find another way of specifying the elements that are the members of each of your sets.
   (b) A vector is a quantity that possesses magnitude and direction. So any object which consists of a real number and a specified direction will count as vector.
   (c) Any collection of real numbers will do. For instance, a simple example might be given by the real number 1. This example is the tangent of either of the two equal angles from any isoceles triangle which contains a right angle.
2. (a) The mathematical idea that models the real-world concept of an incline is the idea of a line. When spelling out the mathematical examples that match your real-world examples, make sure that you specify the equation of each line. Note also that the idea of a gradient is relevant in both cases: sometimes the same word is used to describe a mathematical idea and the real-world concept which it models.
   (b) It is easy to imagine examples of the tension in string: you can take out a piece of string and pull on each end with a certain force. But determining the mathematical examples that match each of these particular

real-world examples will be something of a challenge! Perhaps you can think of some situations where you can determine more easily the matching mathematical examples. Note that the relevant mathematical idea that matches the concept of the tension in a string is the idea of a vector.

(c) We can model this concept from the real world with various mathematical ideas, the most relevant of which are the idea of a positive real number and the idea of a positive integer.

Exercise 2c

1. (a) Any irrational number will qualify as a particular instance of this set.
   (b) One of your particular instances should certainly be $x = \sqrt[3]{7}$.
2. We will especially be interested in the particular instances given by the start and the end of the given period of time (along with the matching numbers), but the two points during the given period of time at which the gradient of the curve changes its sign are also of interest.
3. Clearly the second collection of examples conveys the mathematical idea of a set more effectively than the first collection. But even this second collection still fails to do justice to the idea, which can contain any object imaginable rather than just letters, numbers or in the case of the last example nothing at all.

Extension Exercise 2

1. (a) The simplest examples of which I can think are sets which contain just the identity element for a given binary operation, so take the set 0 under the binary operation addition. A typical example might be the set of all integers under the operation addition. And have you found a more unusual example? Make sure that you test each of your examples against the definition of an abelian group.
   (b) A simple example of a linearly independent set of vectors might be given by the following:

   $$\left\{ \begin{pmatrix} 1 \\ 0 \\ 0 \end{pmatrix}, \begin{pmatrix} 0 \\ 1 \\ 0 \end{pmatrix}, \begin{pmatrix} 0 \\ 0 \\ 1 \end{pmatrix} \right\}$$

   It is after all impossible to express any one of either of these vectors as a linear combination of the others. An unusual example might simply involve a set of vectors that have been expressed in an unusual form, perhaps in the representation of a polynomial. The scope is endless.

(c) The real-valued function given by $f(x) = \sin x$ provides an interesting non-example of an injective function. This is because an infinite number of real numbers all have assigned to them the same real number by this function. For instance the real numbers $\ldots, -3\pi, -2\pi, -\pi, 0, \pi, 2\pi, \ldots$ all have assigned to them the real number 0. For an injective function we would need to avoid even two real numbers having the same real number assigned to them.

2. (a) A uniform cast-iron sphere will have its centre of gravity at the centre of the sphere. A uniform plank of wood will have its centre of gravity at a point which is both halfway across length of the plank and halfway along the width of the plank. Note that both of these real-world examples are matched by examples of the mathematical idea of a point; all you need to do to ensure that these mathematical examples are well-defined is to specify in some way the location of each point.

(b) We can model the rate over the given period of time using the idea of function. An unusual example might be the rate for a population which is extinct (which will be 0 over the whole given period of time).

▶ Chapter 3

Exercise 3a

1. A trapezium consists of four straight lines. The end of each line is joined to the end of another line to form a closed shape with four sides. In addition, one pair of opposite sides are parallel.
2. Take the growth of a tree or walking around a maze, or an experiment which involves tossing a coin to determine which of alternative courses of action to take. It is surprising how wide a range of situations in the real world can be modelled by the same simple visual idea.

Exercise 3b

1. Either option will work with this one.
2. It probably makes more sense to start with the overall picture in this case, but then this example is also sufficiently simple to start with the details.
3. I would certainly want to start this case by filling in some details!

Exercise 3c

1. If you draw the graph of the function you should be able to see the following. As far as an overall impression is concerned, it is interesting to note

that the curve increases ever more steeply as the value of x increases between 0 and 2. If we link this to our definition of the integral, we can observe that the part of our specified region close to 2 provides a much greater proportion of the area of the region than the part of the region close to 0, and this yields insight into the relationship between differentiation and integration (between gradient and area). When a curve increases more steeply as x increases then the area under that curve will increase steeply (even more steeply in fact) as the upper limit of integration increases.

If we then focus on a couple of details, we can pick out the points on the graph of the function given by $(0,0)$ and $(2,4)$. Linking these with the definition of the integral we can note that when $x = 0$ the distance between the x-axis and the curve is 0, while when $x = 2$ the distance between the curve and the x-axis is now 4. This helps us to appreciate that the contribution to the area of the specified region is greater where value of the function is greater.

2. My overall impression is that as time increases the acceleration of the body gets closer and closer to the value c. More formally we can say that as $t \to \infty$ then $a \to c$. The only obvious detail to pick out is that when $t = 0$ then $a = 0$. If we express a as a function of t we can say that $a(0) = 0$. As far as a real-world interpretation is concerned, as time increases the body gets closer and closer to accelerating at the given constant rate. So while the body starts by travelling at a constant velocity, it eventually increases its velocity at a constant rate.

Extension Exercise 3

1. There is of course a huge range of possible answers to this question. Part of the purpose in asking it is to allow you freedom to make connections for yourself. You might, for instance, have added a line to your plane and this could be described in formal mathematical terms. Or you might have added a point and this again can be described in mathematical terms.
2. We can link this diagram to Newton's law of gravitation: that two bodies attract each other with a force that is inversely proportional to the square of the distance between them. And in this case the real-world concepts involved are the centre of gravity of a body, the distance between two objects and the force acting on a body, in this case denoted by **F**.
3. We can of course say that when $x = a$ then $f(a) = c$. But of greater interest is the observation that the derivative of the function at a is given by the gradient of the line.

▶ Chapter 4

Exercise 4a

1. It is guaranteed to serve some purpose!
2. (a) These symbols do not refer to anything in particular. As it stands they have no meaning, so it is not surprising that they will make little sense to you.
   (b) Hopefully the impression these symbols make on you is not the same impression that the mathematical symbols with which you deal on a regular basis make on you!

Exercise 4b

1. If the 'official' definitions differ from your own definition, can you account for the difference? Is your definition still correct?
2. Only the usage of the symbol **R** in (b) is acceptable. A literal translation of the text in (a) would be 'two is an the set of all real numbers' – which makes no sense at all. And in (c) **R** is not allowed to stand for the word 'real': it is only allowed to stand for the set specified.

Exercise 4c

1. In formal terms this inequality asserts that if you start with the expression given by twice the real variable x minus 3, and take away from this the real variable x, then the resulting expression is a member of the set of all negative real numbers. The solution to this inequality is of course '$x < 3$' whose meaning is that the real variable x minus three belongs to the set of all negative real numbers.
2. The vector which represents the force acting on a particle is equal to the scalar quantity which represents the mass of the particle multiplied by the vector, each of whose components is given by the derivative of the function which specifies the corresponding component of the velocity of the particle. (Note: a more complete translation would also require us to spell out the ideas a vector and the derivative of a function.)

Extension Exercise 4

1. (a) In particular, the text $A \subseteq B$ means that every member of the set A is also a member of the set B, the symbol $\Leftrightarrow$ means more than 'is equivalent to' (as we shall see in Chapter 6) and $A \cap B$ refers to the set which contains the elements which belong to both of the sets A and B.
   (b) More is needed than the text 'a cross b equals a b sine theta n hat'! What do each of these terms actually represent? And, of course, what

happened to the process of multiplication which occurs three times on the right-hand side of this equation?
2. The original differential equation means the following. The rate at which the size, y, of the population is increasing at any moment of time is equal to the size, y, of the population at that same moment of time. The solution of the differential equation is that $y = ae^t$, where a is a constant. This means that as time increases, the size of the population increases exponentially.

▶ Chapter 5

Exercise 5a

1. You are unlikely to understand the ideas 'alphabet' and 'word'. Failure to understand any of the simpler ideas involved in a new idea will make it difficult to understand the new idea.
2. The identity is $\cos x - \cos y = -2 \sin\left(\frac{x-y}{2}\right) \sin\left(\frac{x+y}{2}\right)$, although it is only too easy to forget.

Exercise 5b

1. (a) Vector quantity; process of multiplication; particle; mass; and velocity.
   (b) Positive real number; variable; power; equation; and the irrational number $e = 2.718\,281\,8\ldots$
2. (a) Function; the relationship between two objects of an assignment; real number; variable; angle; size or measure; radians; and the tangent of a real number.
   (b) Right angle; positive; and radians.

Exercise 5c

1. Ideas from logic: statement; a statement is true; a statement is false. Other ideas: set; ordered pair of elements; and element from a set.
2. Real-valued function; real number; constant; area; region; Cartesian grid; boundary; graph of a function; line; equation; variable; and axis.
3. The idea of a power has been omitted. We could also include the idea that the operations of arithmetic are performed in a specific order.

Extension Exercise 5

1. (a) The idea of a group depends on the ideas of a non-empty set, a binary operation and axiom. The four axioms depend on the following ideas.

Closure: member of a set, binary operation, and non-empty set. *Associativity*: member of a set, binary operation, non-empty set, binary operation, order of operation, and equality. *Identity*: member of a set, non-empty set, binary operation and equality. *Inverse*: member of a set, non-empty set, binary operation, equality, and identity element.

We can see how complicated the idea of a group is by looking at this above analysis. The ideas which are involved in the idea of a group are all closely interrelated with each other. For instance, the idea of a group depends on the ideas of a non-empty set, a binary operation and an axiom. But each of the four axioms also depends on the idea of a non-empty set and a binary operation.

(b) The idea of a continuous function depends on the following ideas: function; point; constant; domain of a function; the logical condition 'if and only if'; variable; positive real number; modulus of a real number; inequality; implication; and image of a member of the domain.

2. The relevant ideas are as follows: the idea of a reactant, the concentration of a reactant, a function (in this case $c(t)$), the derivative of a function, a variable, a constant, an equation and the notion of negation.

▶ Chapter 6

Exercise 6a

1. If you try enough points you will find that the pattern breaks down eventually; specifically when $n = 6$. Guessing patterns is sometimes useful in mathematics but is never sufficient.
2. Intuition might indicate that $0.\dot{9}$ is a number slightly less than one. But if it is slightly less than 1 you ought to be able to indicate how much less than 1 it is, and this cannot be done. In fact, $0.\dot{9} = 1$.
3. It is surprising how often in everyday life the meaning of a word changes when the context changes. In the first of the two sentences there is an evident connection between the first part of the sentence and second part of the sentence, while in the second of the two sentences there is no evident connection between the two parts of the sentence.

Exercise 6b

1. (a) This sentence is a statement, a false statement.
   (b) All we have here is an object. It can be neither true nor false and hence it is not a statement.
2. (a) True.

(b) This statement claims that π is a rational number, which is of course not true.

3. (a) Not valid: there are isoceles triangles which are not equilateral. But by contrast it is true to say that if T is an equilateral triangle then T is also an isoceles triangle.

(b) Valid: it is true to say that if x is less then 1 then it must be less than 2 as well. Indeed, this is true for every real number x.

Exercise 6c

1. (a) True.
   (b) False.
   (c) True.
2. (a) Not valid: what happened to the constant of integration?
   (b) Not valid: if we substitute, for instance, the value $x = 2.5$ into both of these statements, then the first statement ($x < 2$) is false and the second statement ($x < 3$) is true; so these two statements cannot be equivalent to each other.

Extension Exercise 6

1. (a) The first two equations are not equivalent to each other. And it is just as well that a logical mistake has been made because if you substitute either of the values $x = 1$ and $x = -4$ into the first equation then the equation is not satisfied.
   (b) This argument is acceptable up to the final sentence. But knowing that the statement $(\sin^2 x / \cos^2 x) + 1 = 1/\cos^2 x$ is true does not tell us whether the statement $\tan^2 x + 1 = \sec^2 x$ is true. All we know is that *if* the statement $\tan^2 x + 1 = \sec^2 x$ is true then the statement $(\sin^2 x / \cos^2 x) + 1 = 1/\cos^2 x$ is true.
2. All of these phrases are used between two statements, and each phrase indicates that when either one of the statements is true the other statement is true, and that when either one of the statements is false the other statement is false.

▶ Chapter 7

Exercise 7a

1. There is of course very little to link the words in the first set with each other. By contrast, the words in the second set are all connected to each other according to a familiar pattern.

Exercise 7b

1. The nth term in the sequence is given by $1 + 2(n^2 - 1)$. One way of spotting this is to notice how close this sequence is to a sequence of square numbers. Alternatively it is possible to consider the difference between each term in the sequence as the sum of an arithmetic progression with starting term 6 and difference 4.
2. These data provide examples of Newton's inverse square law of gravitation.
3. None of the letters s, d and t has been defined. Furthermore, the letter s has been used to mean two different things.

Extension Exercise 7

1. (a) Such ideas as subset, proper subset, complement, union and intersection form important types of connection in this area, so look out for these.
   (b) You will find it useful to refer back to some of the fundamental ideas on which this area of probability (as well as statistics) depends. The idea of a discrete random variable itself is of course important, but it will also be important to link your work to the main axioms on which all of probability depends.

▶ Chapter 8

Exercise 8a

1. Measuring the angle will of course not provide an exact solution. Another option is just to draw the circle with the relevant angles. You might then try to construct some triangles and hope this leads to a solution. Alternatively, you can see whether any theorems already exist that link the ideas contained in the statement of the problem.
2. We will need a result that links the unknown force to information that we either already have or can find out on the basis of other information given in the problem.
3. Monitoring strategies that you might have employed would include checking for mistakes, analysing the way you tackled the problem, awareness that a plan of attack was not apparently leading towards a solution and awareness that the solution was not becoming unrealistically complicated.

Exercise 8b

1. This problem is intended to be relatively simple. Rather than providing a solution for you, it is important here to point out that your commentary

should be significantly longer than your solution. Have you indicated why you made each of the steps that you took? Can you note how each of your steps fits into the structured approach to problem-solving that has been advocated in this chapter?
2. You will need to draw on your wider knowledge of triangles in order to solve this problem. But again, the focus of your work should be on adding a thorough commentary.

Extension Exercise 8

1. If you solve enough problems and fully characterise the strategies you employ, you may well be able to see how certain types of strategy work well across the area of mathematics or one of its applications that you chose.
2. In solving problems like this one, it is essential to make use of the definition of a group. It is also important to translate information about the order of x and y into a representation involving equations. You should be able to justify every step of your solution in terms of information given in the problem, the definition of a group and other results or definitions from group theory.

▶ Chapter 9

Exercise 9a

1. An actual incline in the real world will not be completely even, so modelling it with a line will only be an approximation of reality. Strictly speaking, a mathematical line extends to infinity in both directions, but if we model the incline with a segment of a line we presume that the incline has a clear beginning and end which might well not be the case. It is also the case that a line is a one-dimensional object (it has no width at all), whereas an incline in the real world will undoubtedly be of a certain width. There may well be more assumptions that are involved in this model. Can you add to those given above?
2. It is, for instance, possible that Euclid thought about the shafts of light that you sometimes see streaming through the clouds. Once someone has set up a good model of some aspect of the real world then the model seems obvious, but for the person trying to model something new a real challenge is usually entailed.
3. The mathematical relationship specified is that ρ is equal to m divided by V. We could interpret this to say that the density of an object is defined as the quantity of mass that is packed into a unit of volume of the object, taking one specific volume as a standard unit of volume.

4. The limitations to this model are seemingly endless. People do not have standard sized legs or steps, so different people will come up with different measurements, and there is no guarantee that the person concerned will walk in a straight line. The model is hardly very reliable.

Exercise 9b

1. (a) The velocity of a body refers to both the speed at which the body is travelling and the direction in which the body is travelling. In considering this concept we are thus already at stage 2 of our modelling cycle. But why do you think someone might have invented the real-world concept of velocity to start with? To cover stage 3 of the modelling cycle we introduce a mathematical idea to model the concept of velocity: the idea of a vector. When trying to solve problems involving the concept of the velocity of a body, this model will thus allow us to draw on all the mathematical theory related to vectors.
   (b) Fortunately someone else has done the hard work for us by introducing this concept already. The most common of the models for this concept is the mean. The mean of a set of data is a precisely defined statistical model of central tendency, even if the term is used more informally in everyday life. But why does the statistical definition of the mean of a set of data provide a good model of central tendency?
   (c) Most models of population growth involve differential equations. Why is this the case?

Extension Exercise 9

1. (a) The mathematics involved in most models of demand and supply is relatively simple, but can you make explicit the connection between the mathematics and the real-world concepts involved?
   (b) Previous scientists had made various observations about the nature of planetary motion. Why was it so important for Newton to explain these observations in mathematical terms?
   (c) The Poisson model draws on a number of concepts from the theory of probability. Have you provided a full explanation of all of these concepts?
2. (a) It is clearly important to note that as the light intensity changes the rate of photosynthesis also changes. How does a mathematical function model this relationship?
   (b) Your answer will need to include the mathematical idea of a function, as this idea underlies this definition of marginal revenue.
3. (a) While this equation will be of most interest to students of chemistry, other students may benefit from trying to see how unfamiliar concepts relate to each other in a mathematical fashion.

(b) Hooke's Law relates the tension in a string to the natural length of the string, to the modulus of elasticity for the string and to the distance the string is extended or compressed. It indicates, for instance, that if the length of the string is increased then the tension in the string will decrease.

▶ Chapter 10

Exercise 10a

1. Choose specific values for x and y and check whether the equation holds. Employ a computer algebra system to plot a graph of the curves $f(x, y) = \log_e(x) + \log_e(y)$ and $g(x, y) = \log_e(x \times y)$. What is the definition of $\log_e$? What connection does this theorem have with the rules for powers of real numbers?
2. Even without understanding all the ideas contained in the statement of this theorem, it is still possible to conclude from the theorem that the distinct equivalence classes of P partition X.
3. Pythagoras' Theorem finds countless applications in mathematical models of the real world, but it is also employed in proving other theorems, such as in the well-known equation linking the sine and cosine functions, as given in Example 6.2.

Exercise 10b

1. This proof is structured around the idea of equivalence. The first four lines of the proof demonstrate that statements (1) and (4) are equivalent to each other. Statement (5) then merely points this out. In summary we can say that the proof involves determining the solutions of the equation $a^2 = b^2$.
2. Statement (1) merely defines two new variables, a and b, while statement (2) provides an equivalent definition of these variables. Statement (2), which we know is true, then establishes the truth of statement (3) by implication. However, statement (3) is equivalent to line (6), which is the theorem. And since statement (6) is equivalent to a statement (3) which we know is true, the statement (6) – and hence the theorem – must also be true. The strategy that underlies this proof is to start with a statement which we know is true, and demonstrate that our theorem is equivalent to this statement. In this case we start with something we know how to deal with (equations involving the constant e).

Exercise 10c

1. (a) In order to move from line (2) to line (3) we need to draw on the laws of arithmetic.
   (b) Because of the Fundamental Theorem of Algebra.
   (c) Line (3) provides one of the links in the chain that enables us to claim that lines (1) and (4) are equivalent to each other.
2. (a) Line (3) draws on the definition of $\log_e$.
   (b) What is needed is a close look at the definition of $\log_e$.
   (c) This is the statement which we both know is true and know how to deal with.
3. (a) In order to understand the sine rule it will help to check whether it does hold in a few specific cases. A diagram of a typical case will certainly help. It will also be important to make sure that you fully understand the sine of an angle and the notation used to represent this idea.

Extension Exercise 10

1. (a) Examples of the theorem, a typical visualisation of the theorem, understanding of the symbols involved, awareness of the ideas that contribute to the statement of the theorem, an understanding of the logic involved and attempts to link the theorem to more familiar mathematics will all be relevant.
   (b) Here it is essential that you fully understand the ideas that are involved in the theorem. For instance, you certainly need to understand what a continuous random variable actually is, but examples and a visual image are also particularly important. It will also help if you can relate the theorem to your own experience in the real world.
2. (a) The Prime Factorisation Theorem states that every integer greater than 1 can be expressed as a product of primes. It will be important to make explicit all of the logical ideas involved in the proof that you have found of this theorem. It might well be the case, for instance, that the idea of implication is central to the proof you found of this theorem and, in terms of an intuitive overview, it might be the case that your proof actually involves outlining a way to create a product of primes for any given integer greater than 1. The comments that you add to your proof can point out how each line links to your intuitive overview or to the logical structure of the proof.
   (b) You will find a variety of proofs of this theorem, but the proof you find may well unpack the term 'bijective' into the categories 'surjective' and 'injective'. After all, a bijective function is a function which is both surjective and injective. We would then have a situation in which the nature of an idea in the theorem shapes the nature of the proof.

Chapter 11

Exercise 11a

1. Even better, make annotating your lecture notes a regular habit.
2. The aim behind this exercise is for you to learn to take the initiative with your tutor.
3. It will help if your questions stem from your own experience of problem-solving.

Exercise 11b

1. How well you tackled the task may well be related to how well you got on as a group.
2. It is surprising how much difference it can make to your own understanding when you try to explain a problem or idea to someone else.

Exercise 11c

1. By now you should be used to employing the skills of Part I in a search for understanding. These skills underpin a great deal of mathematical thinking.
2. The strategies considered here are a common feature of the approach recommended in this book: to focus on both the overview and on the details.
3. You might work with a friend, make use of a computer algebra system or find a new textbook.

Chapter 12

Exercise 12a

1. (a) The more explicitly you are able to do this, the better.
   (b) Can you account for the way in which the output changes as the input is varied?
   (c) It is important to make sure you can fully translate the symbols used in both the input and the output.
2. You may wish to revisit Chapter 3 for guidance on this exercise.

Exercise 12b

1. As well as using the CAS to solve the differential equation and to produce several visual images, try both integrating the expression $\cos x\, e^x$ and differentiating a solution of the differential equation.

184 Answers and Comments for Selected Exercises

2. You may well have found that your work was more effective for the second of these two assignments. Even if this was not the case, can you indicate what role, if any, technology played in the way in which your work for these two assignments differed?

Extension Exercise 12

1. You should be able to see clearly that the function g assigns to the real numbers 0, π and 2π the same real number.
2. It is worth acquiring the skill of being able to learn a new package, given the speed at which software is updated and invented.

▶ Chapter 13

Exercise 13a

1. Clearly the second of these two lines of text is more coherent and clear, but why? It is true that there is some coherence in the first text, in that we can recognise three inequalities. But this first text has not spelled out any connection between the inequalities. By contrast, it is clear in the second text that these three inequalities are all equivalent to each other. And in the first text, what is x? This is not clear. But in the second text it is clear that x is an unspecified real number.
2. We could spell out exactly what **F**, m and **a** all represent. And in addition we could specify the units of force and mass which we are employing.
3. The text may be more suitably written as follows:

$$\text{If } y = a^x, \text{ then } \log_e y = \log_e a^x = x \log_e a$$

$$\text{Hence } \frac{d(\log_e y)}{dx} = \frac{d(x \log_e a)}{dx} \Rightarrow \frac{1}{y}\frac{dy}{dx} = \log_e a$$

$$\text{Hence } \frac{dy}{dx} = a^x \log_e a$$

Exercise 13b

1. Taking responsibility for your own study habits is an important part of success.
2. Carrying out this activity for all of the topics your courses cover will ensure that your revision is as focused as possible.

Bibliography

The publications listed below are those which have either been referred to in the text or have influenced the author in the writing of the text. Books which are particularly suitable for further reading by students are marked with an asterisk.

Allenby, R. B. J. T. *Numbers and Proofs* (London: Arnold, 1997).*
Berry, J., Burghes, D., Huntley, I., James, D. and Moscardini, A. *Teaching and Applying Mathematical Modelling* (Chichester: John Wiley, 1984).
Berry, J. and Houston, K. *Mathematical Modelling* (London: Edward Arnold, 1995).*
Boyer, C. B. and Merzbach, U. C. *A History of Mathematics*, 2nd edn (Chichester: Wiley, 1989).*
Burton, L. and Jaworski, B. *Technology in Mathematics Teaching* (Bromley: Chartwell-Bratt, 1995).
Crawford, K., Gordon, S., Nicholas, J. and Prosser, M. *Learning and Instruction*, 4 (1994) 331–45.
Devlin, K. *Mathematics: The Science of Patterns* (New York: Scientific American Library, 1997).*
Eccles, P. J. *An Introduction to Mathematical Reasoning* (Cambridge: Cambridge University Press, 1997).*
Exner, G. R. *An Accompaniment to Higher Mathematics* (New York: Springer, 1996).*
Leron, U. 'Structuring mathematical proofs', *American Mathematical Monthly*, 90 (1983) 174–85.
Mason, J. *Thinking Mathematically* (Wokingham: Addison-Wesley, 1985).*
Mason, J. *Learning and Doing Mathematics*, 2nd edn (York: QED, 1999).*
McKim, R. H. *Thinking Visually* (Palo Alto: Dale Seymour, 1980).
Northedge, A., Thomas, J., Lane, A. and Peasgood, A. *The Sciences Good Study Guide* (Milton Keynes: The Open University, 1997).*
Pimm, D. *Speaking Mathematically* (London: Routledge, 1987).
Polya, G. *How to Solve it*, 2nd edn (Harmondsworth: Penguin, 1957).*
Schiffer, M. M. and Bowden, L. *The Role of Mathematics in Science* (Washington, DC: The Mathematical Association of America, 1984).*
Schoenfield, A. H. *Mathematical Problem Solving* (San Diego: Academic Press, 1985).

Skemp, R. *The Psychology of Learning Mathematics* (Harmondsworth: Penguin, 1971).*

Smith, R. M. *Mastering Mathematics* (Pacific Grove: Brooks/Cole, 1998).*

Stewart, I. and Tall, D. *The Foundations of Mathematics* (Oxford: Oxford University Press, 1977).*

Tall, D. (Ed) *Advanced Mathematical Thinking* (London: Kluwer, 1991).

Index

abstract algebra
  examples from   21, 49, 118, 126
  see also under individual topics
algebra
  connections within   66
  developing fluency   xiv, 133, 166–169
  examples from   34, 39, 43–5, 57, 61,114–20, 122
analysis
  importance of   4
  of a definition   12–15, 43–5, 47–8
  of an idea   42–9
  of a proof   112–13
  of a text   137
  of a visual image   27–8
  use in problem-solving   81
  use in understanding a proof   121, 123–4
  use in understanding a theorem   113, 116
analysis, examples from   50
applications of mathematics
  see under individual disciplines; and see also modelling
approaches to study   xiii, 3–5, 137,138
assessment
  accuracy   155, 157, 159, 160
  clarity   154, 155, 157–8, 159, 160
  coherence   155, 156–7, 159, 160
  making revisions   154–6, 158, 159, 160
  see also under individual type

attitudes to mathematics and its applications   1–3
axioms   54

biology, applications of mathematics in   20, 21, 22, 91–2, 110, 148–52
business studies, applications of mathematics in   *see* economics

calculus, examples from   *see under individual topics*
chemistry, applications of mathematics in   45–6, 51, 106, 110
combinatorics, examples from   89–91
computer algebra system (CAS)   140–52
connections
  assistance of technology with   145
  importance of   2–5, 63–5
  spotting   66–9
  spelling out   69–71
  use in modelling   101
  use in problem-solving   81, 82
  use in understanding a proof   121
  use in understanding a theorem   114
coursework   156–8
Crick, Francis   22

deductive reasoning   53–5, 57–8

Index

definitions
  analysing *see* analysis
  importance of 12–13, 35
Descartes, René 64
diagrams *see* visual images
differential equations, examples of 40, 62, 148–52

economics, applications of mathematics in 17, 99–105, 110, 157
Einstein, Albert 32
engineering, applications of mathematics in *see* mechanics
equivalent statements 60–2, 116, 122
essays 159
Euclid of Alexandria 54, 106
examinations
  avoiding mental blocks in 163
  preparing for 162–3
  taking 163–4
examples
  assistance of technology with 143
  creating a varied set of 10–17
  simple 15–16, 17
  typical 16–17
  unusual 16–17
  use in modelling 100
  use in problem-solving 81
  use in understanding a proof 121
  use in understanding a theorem 113–14
  see also non-examples
examples classes 135–6

feedback 161
functions
  examples of 12–14, 15, 16, 17–18, 28–9, 93–4
  real-valued 13, 47
  sketching graphs of 25–6

generalisation 67–70, 114
geometry, examples from 23, 31, 58, 64, 87, 95, 113

graphical calculators 140
graphs *see* visual images
group theory, examples from 21, 50, 95
groupwork 136–7

handouts 137

images *see* visual images
implication 36, 57–59, 60–2
inductive reasoning 53–5
information and communication technology *see* technology
integrals, examples of 30, 35, 37, 38, 49
irrational numbers 9, 20, 59, 64, 123–6

lectures 132–4
linear algebra
  connections within 73–4
  examples from 21, 51, 73–4
logic
  assistance of technology with 144–5
  role in course work 156–7
  role in underpinning mathematics 2, 52–5
  role in understanding a proof 119–22
  role in understanding a theorem 116
  see also deductive reasoning

Mason, John 69
mathematical model *see* models
mathematical text
  writing 156–8, 161–2
  translating 37–8
mechanics, examples from 15, 21, 24, 26, 30, 31, 37–8, 46, 69, 71, 88–9, 106–10
memorising facts xiii, 3–5, 21, 64–5, 112
models 10–11, 14, 15, 20, 22, 29–30, 36, 45–6, 66–9, 96–111
  creating 99–104
  interpreting 103–4, 147

limitations of 105–6
uses of 96–7, 105–6
validating 104, 147
modelling
   assistance of technology
      with 146–8
   cycle 99
   data-driven 102
   theory-driven 101

Newton, Issac 46, 55, 100
non-examples 13–14
notation see symbols

oral presentations see
   presentations

parameter 115–16
particular instances of
      examples 17–21, 67–8, 69
   assistance of technology
      with 143, 145, 149
   creating a varied set of 18
   for applications of
      mathematics 19, 20
   use in problem-solving 81
   use in understanding a
      theorem 116–17
physics
   applications of mathematics
      in 10–11, 12, 32, 41–2, 106
   see also mechanics
Picasso, Pablo 27
Pimm, David 38–9
Polya, George 80
presentations, giving 155,
   160–1
problem-solving
   assistance of technology
      with 146
   carrying out a plan 84–5, 87
   devising a plan 82–4, 87
   when revising 163
   in examinations 164
   reviewing your work 85–7
   role of confidence 84, 92
   structured approach 79–80
   understanding the
      problem 80, 82, 86

problems classes see examples
   classes
projects 159
proof
   assistance of technology
      with 148
   by equivalence 60–2
   by implication 67–8
   by contradiction 123–6
   commentary on 121–3
   details of 121–3, 125–6
   structure of 118–21, 122, 125

quantifiers 60

rational numbers 9, 64, 123
real numbers 9
   see also set of all real
      numbers
real-valued function see
   function
real world see models
Recorde, Robert 35
reflection xv
reports 159
revision
   before an examination 162–3
   before a presentation 160–1
   of a text 158
   to understand advanced
      concepts 49–50

set theory
   examples from 17, 21, 36,
      40, 74
   set of all real numbers 18, 36
shapes see visual images
Skemp, Richard 48
skills
   assistance of technology
      with 142
   importance of developing xiii
   use in examples classes 135
   use in lectures 133
   use in reading 137
   use in tutorials 134–135
   see also under individual skills
specialisation see particular
   instances

statement 44, 56–57, 59, 113, 117, 125
statistics, examples from 74, 110, 126, 143
synthesis
  importance of 4
  of a proof 112–13
  of a text 137
  of a visual image 23
symbols
  assistance of technology with 144
  interpreting 33, 34–6
  purpose of 32–4
  role in modelling 69, 101
  role in problem-solving 81
  role in understanding a proof 121
  role in understanding a theorem 115–16
  use in calculations 38–9

tangent of an angle 17, 47
teamwork see groupwork
technology
  mastering 141
  use of 141–2
  see under individual skills and tasks
text see mathematical text
textbooks 137
theorems
  examples of 113, 123
  importance of understanding 113
  uses of 82, 117–18, 124

trigonometry
  connections within 71–72
  examples from 43, 46, 62, 123
  see also under individual topics
truth
  establishing 53–4, 112
  see also statement
tutor 1, 85, 131–5, 155
tutorials 134–5

vectors
  definition 73
  examples of 17, 46, 69, 73–4
variables 41, 48–9, 56–7, 67, 116
visual images
  assistance of technology with 143–4, 149
  details of 27–8, 115
  drawing 23, 24–6
  importance of 22
  looking at 22, 27–8
  putting to good use 22–3, 28–30
  structure of 24–5, 27–8, 115
  use in modelling 28–30
  use in problem-solving 81
  use in understanding a proof 121
  use in understanding a theorem 114–15

Watson, James 22